# 3rd Karl Schwarzschild Meeting (KSM 2017)



## Gravity and the Gauge/Gravity Correspondence

Journal of Physics: Conference Series Volume 942

Frankfurt, Germany
24 - 28 July 2017

**Editors:**

**Piero Nicolini**
**Matthias Kaminski**
**Jonas Mureika**
**Marcus Bleicher**

ISBN: 978-1-5108-5520-5
ISSN: 1742-6588

**Printed from e-media with permission by:**

Curran Associates, Inc.
57 Morehouse Lane
Red Hook, NY 12571

**Some format issues inherent in the e-media version may also appear in this print version.**



Printed by Curran Associates, Inc. (2018)

For permission requests, please contact the Institute of Physics at the address below.

Institute of Physics
Dirac House, Temple Back
Bristol BS1 6BE UK

Phone: 44 1 17 929 7481
Fax: 44 1 17 920 0979

techtracking@iop.org

**Additional copies of this publication are available from:**

Curran Associates, Inc.
57 Morehouse Lane
Red Hook, NY 12571 USA
Phone: 845-758-0400
Fax: 845-758-2633
Email: curran@proceedings.com
Web: www.proceedings.com

# Table of contents

## Volume 942

**3rd Karl Schwarzschild Meeting - Gravity and the Gauge/Gravity Correspondence**

**24–28 July 2017, Frankfurt, Germany**

**Accepted papers received: 15 December 2017**
**Published online: 3 January 2018**

**Contributed Paper**

IOP Conf. Series: Journal of Physics: Conf. Series **942** (2017) 011001 doi:10.1088/1742-6596/942/1/011001

# 3rd Karl Schwarzschild Meeting – Gravity and the Gauge/Gravity Correspondence, 24–28 July 2017, Frankfurt am Main, Germany

**Piero Nicolini**
Frankfurt Institute for Advanced Studies (FIAS), Frankfurt am Main, Germany
Institut für Theoretische Physik, Goethe-Universität Frankfurt am Main, Frankfurt am Main, Germany
E-mail: nicolini@fias.uni-frankfurt.de

**Matthias Kaminski**
Department of Physics and Astronomy, University of Alabama, Tuscaloosa, AL 35487, USA
E-mail: mski@ua.edu

**Jonas Mureika**
Department of Physics, Loyola Marymount University, Los Angeles, CA 90045, USA
E-mail: jmureika@lmu.edu

**Marcus Bleicher**
Frankfurt Institute for Advanced Studies (FIAS), Frankfurt am Main, Germany
Institut für Theoretische Physik, Goethe-Universität Frankfurt am Main, Frankfurt am Main, Germany
E-mail: bleicher@fias.uni-frankfurt.de

## 1. Preface

The Karl Schwarzschild Meeting 2017 (KSM2017) has been the third instalment of the conference dedicated to the great Frankfurter scientist, who derived the first black hole solution of Einstein's equations about 100 years ago.

The event has been a 5 day meeting in the field of black holes, AdS/CFT correspondence and gravitational physics. Like the two previous instalments, the conference continued to attract a stellar ensemble of participants from the world's most renowned institutions. The core of the meeting has been a series of invited talks from eminent experts (keynote speakers) as well as the presence of plenary research talks by students and junior speakers.
Specifically the KSM2017 consisted of six kinds of presentations:

1 Karl Schwarzschild lecture

1 panel discussion

1 introductory lecture (30 minutes long)

IOP Conf. Series: Journal of Physics: Conf. Series **942** (2017) 011001 doi:10.1088/1742-6596/942/1/011001

10 keynotes (60 minutes long)

43 contributed talks (20 minutes long)

14 poster presentations.

Students, junior and senior contributed speakers were selected among participants on the basis of their CV, publication record and abstract of the proposed contribution.

Talks were grouped together by theme as much as possible in five thematic sessions that have been scheduled according to the following scheme:

Holography/Information/Analogue Gravity - I (Monday)

Black hole and Astrophysics - I (Tuesday)

Holography/Information/Analogue Gravity - II (Wednesday)

Black holes theory (Thursday)

Black hole and Astrophysics - II (Friday).

The Karl Schwarzschild Lecture was delivered by Juan Maldacena (IAS) on the topic of the Schwarzschild solution, followed by a lively question period from the audience. A panel consisting of four observationalists and theorists – Martin Ammon (Jena), Avery Broderick (Waterloo/Perimeter), Juan Maldacena (IAS), Erik Verlinde (Amsterdam) – and moderated by Ruth Gregory (Durham) later in the week sought to assess when and where quantum gravity signals might be realizable. Johanna Erdmenger (Würzburg) delivered a lecture about the Gauge/Gravity Correspondence on an introductory level.

The focus on the younger generation of scientists was again a highlight of the meeting, with a new prize category being introduced. The award for overall winner of the Student-Junior joint session went to Mr. Van-Liem Saw (University of Otago), for his presentation "Mass-loss due to gravitational waves with $\Lambda > 0$". The best student talk award was given to Mr. Michael Wondrak (FIAS, Frankfurt) for his talk "AdS/CFT far from Equilibrium in a Vaidya Setup". Honorable mentions in this category were awarded to Mr. Nikola I. Gushterov (Oxford), Mr. Alexander Gu*β*mann (LMU Munich), and Mr. Luciano Manfredi (LMU Los Angeles). The Schwarzschild Prize for best junior scientist talk was awarded to Dr. Viancheslav Emelyanov (KIT, Karlsruhe), who spoke on "Local quantum effects in QED induced by evaporating black holes". The honourable mention in this category was given to Dr. Mario Flory (Jagiellonian University).

The KSM2017 saw the participation of 123 scientists from 24 countries over 6 continents. This included 78 junior and senior scientists, and 45 students.

We present these proceedings as a tribute to the frontier science presented at our meeting, and look forward to future instalments of the KSM.

Frankfurt am Main, December 2017

*Piero Nicolini*
*Jonas Mureika*
*Matthias Kaminski*
*Marcus Bleicher*

IOP Conf. Series: Journal of Physics: Conf. Series **942** (2017) 011001 doi:10.1088/1742-6596/942/1/011001

## 2. Conference photo and poster

**Figure 1.** Conference photo.

IOP Conf. Series: Journal of Physics: Conf. Series **942** (2017) 011001 doi:10.1088/1742-6596/942/1/011001

**Figure 2.** Conference poster.

IOP Conf. Series: Journal of Physics: Conf. Series **942** (2017) 011001 doi:10.1088/1742-6596/942/1/011001

## 3. Sponsors and funding acknowledgments

The organizers of the KSM2017 are grateful for the generosity of the Frankfurt Institute for Advanced Studies for providing the spectacular venue at which the meeting took place. We thank the Deutsche Forschungsgemeinschaft (DFG), the Helmholtz International Center for FAIR, and the Freunde und Förderer der Goethe-Universität Frankfurt for their financial support, as well as the Physikalischer Verein Frankfurt. Special thanks go to Silke Britzen (MPI Bonn), Sven Köppel (FIAS), Ilia Musco (LUTH Paris/Meudon) who served as committee members for the Karl Schwarzschild Prizes for the best presentations in the student and junior scientist categories as well as the overall best student-junior joint session presentation. For their help and support of our efforts, we extend a special acknowledgment to our International Advisory Committee. We are also grateful to Marco Knipfer (FIAS, Frankfurt) for his technical assistance in various phases of the editing of such proceedings. Finally, we are indebted to all participants who helped to make the KSM2017 a success.

## 4. Committees

**Organizing Committee:**
Piero Nicolini (FIAS and Goethe University, Frankfurt)
Matthias Kaminski (University of Alabama, Tuscaloosa)
Jonas Mureika (Loyola Marymount University, Los Angeles)
Marcus Bleicher (FIAS and Goethe University, Frankfurt)

**International Advisory Committee:**
Claus Lämmerzahl (ZARM & University of Bremen)
Robert Mann (University of Waterloo & Perimeter Institute)
Martin Reuter (Johannes-Gutenberg University, Mainz)
Dam T. Son (University of Chicago)
Herman Verlinde (Princeton University)

## 5. List of participants

ALTAŞ KİRACI, Emel, Middle East Technical University, Ankara, Turkey
AMMON, Martin, Friedrich-Schiller University Jena, Jena, Germany
ANDRIOT, David, MPI, AEI , Potsdam-Golm, Germany
APPELS, Michael, Durham University, Durham, United Kingdom
ASTORINO, Marco, UAI, Viña del Mar, Chile
BAMBI, Cosimo, Fudan University, Shanghai, China
BLAZQUEZ SALCEDO, Jose Luis, Oldenburg University, Oldenburg, Germany
BLEICHER, Marcus, FIAS, Frankfurt am Main, Germany
BONANNO, Alfio, INAF - Catania Astrophysical Observatory, Catania, Italy
BOUSSO, Raphael, UC Berkeley, Berkeley, United States
BREU, Cosima, Goethe University Frankfurt am Main, Frankfurt am Main, Germany
BRITZEN, Silke, Max-Planck-Institut für Radioastronomie, Bonn, Germany
BRODERICK, Avery, University of Waterloo / Perimeter Institute for Theoretical Physics, Waterloo, Canada
BUONANNO, Alessandra, Max Planck Institute for Gravitational Physics, Potsdam, Germany
BURINSKII, Alexander, NSI, Russian Academy of Sciences, Moscow, Russian Federation
CABEZAS-RIVAS, Esther, Institut für Mathematik, Goethe Universität Frankfurt, Frankfurt, Germany

IOP Conf. Series: Journal of Physics: Conf. Series **942** (2017) 011001 doi:10.1088/1742-6596/942/1/011001

CADONI, Mariano, Dipartimento di Fisica, Università di Cagliari, Cagliari, Italy
CARR, Bernard, Queen Mary Unversity of London, London, United Kingdom
CASADIO, Roberto, Dipartimento di Fisica e Astronomia, Università di Bologna, Bologna, Italy
CHARZYŃSKI, Szymon, Faculty of Physics, University of Warsaw, Warsaw, Poland
COMPERE, Geoffrey, ULB, Brussels, Belgium
CRUZ-OSORIO, Alejandro, Frankfurt Institute for Advanced Studies, Frankfurt, Germany
CSERNAI, Laszlo, University of Bergen, Bergen, Norway
DAI, De-Chang, Shanghai Jiao Tong University, Shanghai, China
DAUS, Tristan, Goethe University Frankfurt, Frankfurt, Germany
DE LAURENTIS, Mariafelicia, Institute of Theoretical Physics, Goethe University, Frankfurt am Main, Germany
DENG, Gao-Ming, Beijing University of Technology, Beijing, China
EL KHCHINE, Younes, Polydisciplinary Faculty of Taza, Sidi Mohamed Ben Abdellah University, Taza, Morocco
EL-MENOUFI, Basem, U. of Sussex, Brighton, United Kingdom
EMELYANOV, Viacheslav, Karlsruhe Institute of Technology, Karlsruhe, Germany
ERDMENGER, Johanna, Universität Würzburg, Würzburg, Germany
FALCKE, Heino, Department of Astrophysics/IMAPP, Radboud University Nijmegen, Nijmegen, Netherlands
FLORY, Mario, Jagiellonian University, Krakow, Poland
FORINI, Valentina, Humboldt University Berlin, Berlin, Germany
FUJII, Shunsuke, National Institute of Technology, Tokyo College, Hachiohji-City, Tokyo Japan
GALLEGO TORROMÉ, Ricardo, Frankfurt Institute for Advanced Studies, Frankfurt, Germany
GARATTINI, Remo, University of Bergamo, Dalmine (Bergamo), Italy
GARBISO, Markus, The University of Alabama, Tuscaloosa, United States
GARDEMANN, Markus, Friedrich Schiller University Jena, Jena, Germany
GEBHARDT, Charlotte, Goethe University Frankfurt, Frankfurt am Main, Germany
GIUGNO, Andrea, Arnold Sommerfeld Center, LMU Munich, Munich, Germany
GIUSTI, Andrea, Università di Bologna and LMU Munich, Bologna Italy
GONZALEZ-MARTIN, Sergio, IFT UAM/CSIC, Madrid, Spain
GREGORY, Ruth, Durham University, Durham, United Kingdom
GRIENINGER, Sebastian, Friedrich-Schiller University Jena, Jena, Germany
GUSHTEROV, Nikola Ivanov, Oxford University, Oxford, United Kingdom
GUSSMANN, Alexander, LMU Munich, Munich, Germany
HARTMANN, Betti, Instituto de Física de São Carlos, Universidade de São Paulo, São Carlos, Brazil
HOFMANN, Stefan, Arnold Sommerfeld Center for Theoretical Physics, Munich, Germany
JANKOWSKI, Jakub, University of Wroclaw, Wroclaw, Poland
KALISCH, Michael, University of Jena, Jena, Germany
KAMINSKI, Matthias, University of Alabama, Tuscaloosa, United States
KEHM, Dirk, FIAS, Frankfurt, Germany
KIM, Keun-Young, Gwangju Institute of Science and Technology (GIST, Korea), Gwangju, Republic of Korea
KIRKLIN, Josh, DAMTP Cambridge, Cambridge, United Kingdom
KLING, David, Frankfurt am Main, Germany
KNIPFER, Marco, FIAS, Frankfurt am Main, Germany
KOCH-STEINHEIMER, Peter, ITP University Frankfurt, Frankfurt am Main, Germany
KOEGLER, Maximilian, Ludwig-Maximilians-Universitaet Muenchen, Munich, Germany
KOEHLER, Jonas, Goethe University, Frankfurt am Main, Germany

IOP Conf. Series: Journal of Physics: Conf. Series **942** (2017) 011001 doi:10.1088/1742-6596/942/1/011001

KOEPPEL, Sven, FIAS, Frankfurt am Main, Germany
KOIRALA, Roshan, University of Alabama, Tuscaloosa, United States
KONOPLYA, Roman, Sileasean University in Opava, Opava, Czech Republic
KUNZ, Jutta, University of Oldenburg, Oldenburg, Germany
LAEMMERZAHL, Claus, ZARM, University of Bremen, Bremen, Germany
LU, Rusen, Max Planck Institute for Radio Astronomy, Bonn, Germany
MAEDA, Kengo, Shibaura Institute Of Technology, Saitama, Japan
MALDACENA, Juan, Institute for Advanced Study, Princeton, United States
MANFREDI CONSOLE, Luciano, Loyola Marymount University, Los Angeles, United States
MEYER, René, JMU Würzburg, Würzburg, Germany
MIAO, Yan-Gang, Nankai University, Tianjin, China
MIGNEMI, Salvatore, Università di Cagliari, Cagliari, Italy
MIZUNO, Yosuke, ITP, Goethe University Frankfurt, Frankfurt am Main, Germany
MODESTO, Leonardo, Southern University of Science and Technology (SUSTech), Shenzhen, China
MOST, Elias Roland, Goethe University Frankfurt, Frankfurt am Main, Germany
MTOUGUI, Sara, University Mohamed V Faculty of science, Rabat, Morocco
MUREIKA, Jonas, Loyola Marymount University, Los Angeles, United States
MUSCO, Ilia, LUTH (Observatoire de Paris/Meudon), Paris, France
MÜLLER, Fabian, Goethe-University, Frankfurt a. M., Germany
MÜNCH, Johannes, FIAS, Frankfurt, Germany
NATHANAIL, Antonios, Institute for Theoretical Physics, Goethe University, Frankfurt am Main, Germany
NICOLINI, Piero, FIAS, Frankfurt am Main, Germany
O'BANNON, Andrew, University of Southampton, Southampton, United Kingdom
OLIVARES SÁNCHEZ, Héctor Raúl, ITP, Goethe-Universität Frankfurt, Frankfurt, Germany
OLIVERI, Roberto, Université Libre de Bruxelles (ULB), Bruxelles, Belgium
PANOSSO MACEDO, Rodrigo, Queen Mary University of London, London, United Kingdom
PAPENFORT, Ludwig Jens, ITP, Frankfurt, Germany
PERLICK, Volker, ZARM, University of Bremen, Bremen, Germany
PETIT, Jean-Pierre, Pertuis, France
PHAM, Anh Quang, Goethe University Frankfurt, Frankfurt, Germany
PRIBYTOK, Anton, University of Latvia, Riga, Latvia
RACHWAL, Leslaw, Fudan University, Shanghai, China
RADU, Eugen, Aveiro University, Aveiro, Portugal
RATTAY, Michael, Universtität Frankfurt, Frankfurt, Germany
RECHENBERGER, Stefan, GU Frankfurt, Frankfurt, Germany
RENNERT, Julian, University of Waterloo, Waterloo, Canada
SALEK, Abdol Sabor, Goethe-Universität Frankfurt am Main, Frankfurt am Main, Germany
SARKAR, Debajyoti, Ludwig Maximilians University, Munich and ITP Bern, Munich, Germany
SATOW, Daisuke, Frankfurt am Main, Germany
SAW, Vee-Liem, University of Otago, Dunedin, New Zealand
SCARDIGLI, Fabio, Politecnico of Milano, Milan, Italy
SCHRAM, Zsolt, University of Debrecen, Debrecen, Hungary
SEBASTIANI, Lorenzo, Università di Trento, Trento, Italy
SMIT, Roman, Goethe-University Frankfurt/Main, Frankfurt/Main, Germany
STRUCKMEIER, Jürgen, Frankfurt Institute for Advanced Studies (FIAS), Frankfurt am Main, Germany
TANAHASHI, Norihiro, Department of Physics, Osaka University, Osaka, Japan
TRIPATHI, Ashutosh, Fudan University, Shanghai, China

IOP Conf. Series: Journal of Physics: Conf. Series **942** (2017) 011001 doi:10.1088/1742-6596/942/1/011001

TURSUNOV, Arman, Silesian University in Opava, Opava, Czech Republic
TUVERI, Matteo, University of Cagliari and INFN Cagliari, Cagliari, Italy
TZIKAS, Athanasios, Frankfurt Institute for Advanced Studies (FIAS), Frankfurt am Main, Germany
UNRUH, William, UBC, Vancouver, Canada
UZAWA, Kunihito, Kwansei Gakuin University, Sanda, Hyogo, Japan
VERLINDE, Erik, University of Amsterdam, Amsterdam, Netherlands
VERMEULEN, Valery, Erasmus University College Brussels, Ghent, Belgium
VILLHAUER, Elena, University of California, Irvine, Irvine, United States
WEIH, Lukas, Frankfurt, Germany
WITHAM, Oliver, ITP Frankfurt, Frankfurt am Main, Germany
WONDRAK, Michael Florian, Frankfurt Institute for Advanced Studies (FIAS), Frankfurt am Main, Germany
WU, Chao, Wigner Research Center for Physics, Hangarian Academy of Sciences, Budapest, Hungary
WYSTUB, Stephan, Goethe Universität Frankfurt, Frankfurt, Germany
YANG, Run-Qiu, Korea Institute for Advanced Study, Republic of Korea
YOUNSI, Ziri, Goethe Universität Frankfurt, Frankfurt, Germany
ZHANG, Jia-ju, University of Milano-Bicocca, Milan, Italy

IOP Conf. Series: Journal of Physics: Conf. Series **942** (2017) 011002 doi:10.1088/1742-6596/942/1/011002

# Peer review statement

All papers published in this volume of *Journal of Physics: Conference Series* have been peer reviewed through processes administered by the proceedings Editors. Reviews were conducted by expert referees to the professional and scientific standards expected of a proceedings journal published by IOP Publishing.

IOP Conf. Series: Journal of Physics: Conf. Series **942** (2017) 012001 doi:10.1088/1742-6596/942/1/012001

# Imaging black holes: past, present and future

**Heino Falcke**[1,2]

[1] Department of Astrophysics/IMAPP, Radboud University, P.O. Box 9010, 6500 GL Nijmegen, The Netherlands

[2] ASTRON, Dwingeloo, The Netherlands

E-mail: h.falcke@astro.ru.nl

**Abstract.** This paper briefly reviews past, current, and future efforts to image black holes. Black holes seem like mystical objects, but they are an integral part of current astrophysics and are at the center of attempts to unify quantum physics and general relativity. Yet, nobody has ever seen a black hole. What do they look like? Initially, this question seemed more of an academic nature. However, this has changed over the past two decades. Observations and theoretical considerations suggest that the supermassive black hole, Sgr A*, in the center of our Milky Way is surrounded by a compact, foggy emission region radiating at and above 230 GHz. It has been predicted that the event horizon of Sgr A* should cast its shadow onto that emission region, which could be detectable with a global VLBI array of radio telescopes. In contrast to earlier pictures of black holes, that dark feature is not supposed to be due to a hole in the accretion flow, but would represent a true negative image of the event horizon. Currently, the global Event Horizon Telescope consortium is attempting to make such an image. In the future those images could be improved by adding more telescopes to the array, in particular at high sites in Africa. Ultimately, a space array at THz frequencies, the Event Horizon Imager, could produce much more detailed images of black holes. In combination with numerical simulations and precise measurements of the orbits of stars – ideally also of pulsars – these images will allow us to study black holes with unprecedented precision.

## 1. Introduction

When the name black hole was coined in the sixties – allegedly a term picked up by science journalist Ann Ewing and later popularized by John A. Wheeler[1] – it sparked the imagination: not a dull mathematical, abstract object in a theory of gravity but a somewhat mythical, perhaps scary object with an intriguing visual appearance — something seemingly invisible that nonetheless exists. Ever since, there have been many depictions of black holes in the scientific literature, in the arts, and in the public media — perhaps even more so in the latter.

Ironically, the reason why black holes became fashionable was not because they were black, but because they were bright. This was related to the discovery of quasars more than 50 years ago [see, e.g. 57]. Quasars are extremely luminous and simple arguments could show that the deep potential well of black holes could in principle produce that energy.

A falling test mass $m$ at radius $R$ from a black hole mass $M_\bullet$ woul have a Newtonian potential of $E = GM_\bullet m/R$, where $G$ is the gravitational constant. For a constant mass inflow rate $\dot{m}$ the available energy per time $\dot{E}$ can be converted into a luminosity $L$ with an assumed efficiency $\eta$ of order 10% to yield $L \sim \eta\dot{E} = \eta\frac{GM_\bullet}{R}\dot{m} = \frac{1}{2}\eta\left(\frac{R_s}{R}\right)\dot{m}c^2$, where $c$ is the speed of light and $R$ is

[1] http://www.worldwidewords.org/topicalwords/tw-bla1.htm

IOP Conf. Series: Journal of Physics: Conf. Series **942** (2017) 012001 doi:10.1088/1742-6596/942/1/012001

assumed to be of order the Schwarzschild radius, $R_{\rm S} = 2GM_{\bullet}c^{-2}$. For $\dot{m} \sim 1M_{\odot}/{\rm yr}$ the black hole luminosity $L$ can be of order $10^{46}$ erg/sec for black holes some hundred million solar masses – comparable to inferred quasar luminosities.

Not surprisingly, the topic of early research on black holes and active galactic nuclei (AGN) then mainly focused on the overall spectral energy distribution (SED) of luminous AGN. Because of their large distances, imaging a black hole on scales of the event horizon seemed out of reach and more like a purely academic exercise. The only resolvable structure of quasars was seen in their radio emission that later was identified as coming from relativistic plasma jets emerging from near the putative black holes [e.g., 7, 22].

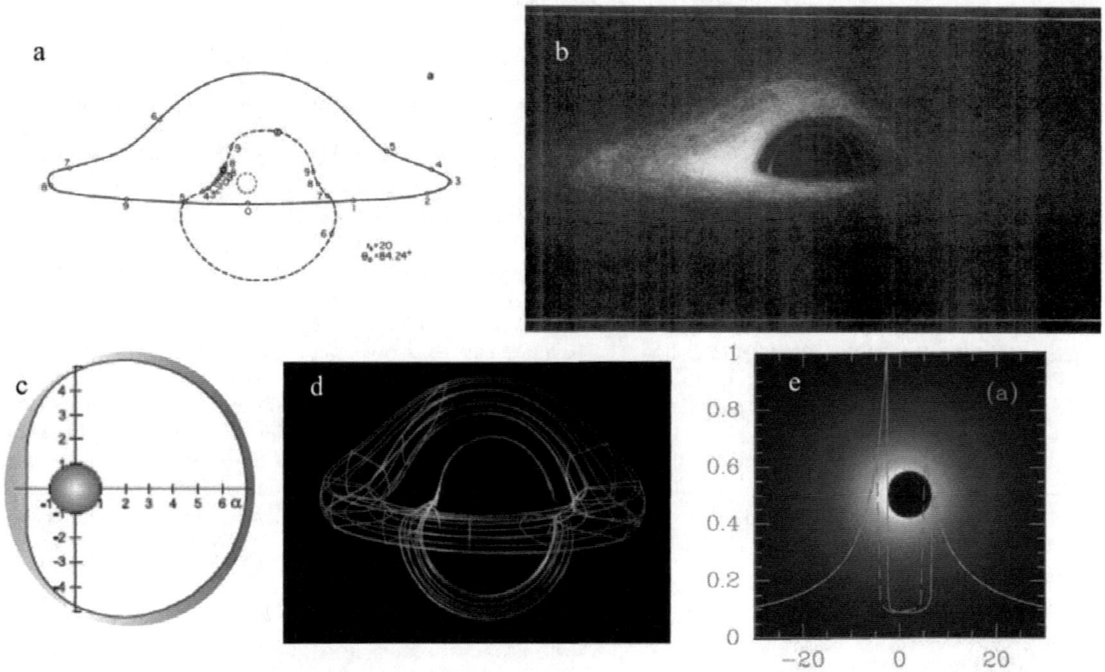


Figure 1 Early representations of black holes. From top left to bottom right: a) star orbiting a black hole [11], b) thin accretion disk extending down to the innermost stable circular orbit (ISCO) [36], c) an extended stellar disk far behind a black hole (Bardeen 1973, as reproduced in Falcke 1999), d) inner edge of an accretion disk seen with color [59], and e) shadow of a black hole surrounded by an optically thin emission region [21].

## 2. Black hole images: early predictions

So, what does a black hole really look like? The first attempt I am aware of[2] dates back to [11], where the authors calculate the appearance of a star in a tight orbit (3 and 20 Schwarzschild radii) around a Kerr black hole (Fig. 1a). The apparent orbit already traces out the basic structures later also seen by more complete accretion disk simulations. However, what we see in these images is not an image of the event horizon at all, it is actually the lensed image of the stellar orbit and true for any gravitating object with or without an event horizon.

The paper that actually caught my attention first on this topic was by [4], published in

[2] Cunningham & Bardeen also refer to Campbell & Matzner [9], doing a similar calculation for a Schwarzschild black hole, but neither motivation nor results seem particularly relevant here.

IOP Conf. Series: Journal of Physics: Conf. Series **942** (2017) 012001 doi:10.1088/1742-6596/942/1/012001

a conference proceedings, where it seemed hardly noticed and cited until then[3]. In this paper Bardeen calculates the appearance of an extended stellar disk behind a Kerr black hole (Fig. 1c). A similar calculation was later presented by [36] for an extended plane wave source at a large distance, e.g., a star in a binary system or simply a large flash light shining at the black hole from behind.

Here an important effect is noticeable, which is related to the photon orbit. The photon orbit is an orbit where light can go around black holes on a closed loop. In practice this means that light entering at or around the photon orbit, will experience a sharp divide: if the light ray is a little bit too close it will disappear behind the event horizon, if it is a little bit outside, it will be able to escape to infinity. The consequence is indeed a sharp image of the event horizon.

In the same paper Luminet also presents his famous image (Fig. 1b). The figure is hailed as the first ever computer-generated image of a black hole. However, as in the case of [11], the dark structure traced out in the center is not actually an image of the black hole itself, but is a lensed image of the inner edge of the accretion disk. Consequently, the image depends crucially on the assumption of where the disk ends. The standard view at the time [49, 50] was that the disk terminated and stopped radiating at the innermost stable circular radius (ISCO, then often called marginally stable radius $r_{\rm ms}$). The same is true for later efforts [59, Fig. 1d]. Strictly speaking we see therefore in these images mainly a hole in an accretion disk rather than the black hole itself.

## 3. The Galactic Center: where dreams come true

A region of high hopes for black hole research had been the center of our own Galaxy [37]. The compact radio source, Sgr A*, found by Balick & Brown [3], was often compared to radio cores in AGN, and hence considered to mark the supermassive black hole in our Milky Way. Near-infrared measurements of stellar orbits [14, 26], revealing stars moving with up to 10,000 km/s around this radio source, and very-long baseline interferometry (VLBI) observations [54], have confirmed that indeed there is a dark mass of about 4.3 million solar masses [27] within a few Schwarzschild radii of Sgr A* [see 25, 20, for reviews][4].

The nature of this radio emission was long under debate [41]. In the early nineties, we developed a model to describe Sgr A* as a starving black hole with a low-power radio jet, fed by a highly sub-Eddington accretion flow [18, 19]. A competing model was to explain all the radio emission from a radiatively inefficient accretion flow (RIAF) [48] at much higher accretion rates. Today we often combine those elements [60, 45] and keep the low accretion rate now demanded by polarization observations of Sgr A* [6, 38].

A prediction of the jet model was that the radio emission should come from closer to the black hole the higher the frequency of the emission was, meaning that the submm-wave bump [42] in the spectrum should come from a compact emission region near the event horizon. One of the first multi-telescope and multi-wavelength campaigns for Sgr A*, involving four telescopes, confirmed that picture [16]. Simple theory also predicted that the emission should become optically thin at and above 230 GHz. Hence, we proposed in the same paper that one should be able to to image the event horizon against this background with submm-wave VLBI observations.

Since the original Bardeen picture only considered emission from far behind the black hole, we then calculated the appearance of a black hole surrounded by an optically thin emission region [21, Fig. 1e], which was more applicable to Sgr A*. Indeed, we found a dark region of about some 5 Schwarzschild radii in diameter, surrounded by a bright ring of emission. That dark region is not just a sharp silhouette as sometimes implied. Interstellar scattering of radio waves and a blend of emission in front of and behind the black hole will give the feature a more

[3] Intriguingly, this is the book, where the famous paper by Novikov & Thorne [49] was published and which has been cited more than 500 times by now.

[4] For parameters of Sgr A* used here, see the overview in Table 1 in Falcke & Markoff (2013).

IOP Conf. Series: Journal of Physics: Conf. Series **942** (2017) 012001 doi:10.1088/1742-6596/942/1/012001

diffuse appearance. Hence, we called it the "shadow" of the event horizon. Many papers now exist that calculate black holes shadows, also in non-standard theories of gravity, suggesting its usefulness for testing GR [see 20, 30, for reviews].

Simulations are now becoming much more sophisticated and are done using numerical general-relativistic magnetohydrodynamic simulations (GRMHD) with ray tracing to take the astrophysical model into account. This was pioneered by [47], initially for a pure disk model, but now also including jet emission [see 44, for a recent review]. The new BHAC code [51] now also allows one to perform these simulations in arbitrary parametrized space times [32]. In Moscibrodzka & Falcke [45] and Moscibrodzka et al. [46] we showed that the original simple analytical jet model does in fact hold up in GRMHD simulations, reproduces the observational characteristics, and produces radio mm-wave emission close to the event horizon — just as needed for shadow imaging.

The diameter of the shadow would be of order $10\,GM_\bullet c^{-2}$, i.e. 63 million km for $M_\bullet = 4.3\times10^6 M_\odot$, corresponding to 51 microarcseconds ($\mu$as) for a distance of 8.3 kpc to the Galactic Center. The resolution of a radio interferometer is given by the ratio of observing wavelength and telescope separation $\lambda/D$. For global interferometers $D$ reaches typical values of 15,000 to 6,000 km, giving one a resolution between 18 to 50 $\mu$as at 230 GHz (i.e., $\lambda$1.3mm) — just large enough to start resolving the shadow. Moreover, the scattering effect decreases with $\lambda^2$, also approaching 22 $\mu$as at $\lambda = 1.3$ mm. Hence, 230 GHz becomes a 'magical' observing frequency, where interstellar scattering, optical depth in the source, global VLBI resolution, and atmospheric transmission make this experiment just possible. At higher frequencies the earth's atmosphere becomes too hostile and at lower frequencies the source is scatter broadened and intransparent. Hence, we can thank God, that the earth is just large enough and at the right location in our Galaxy, to let us to not only see the light, but also the shadow of a black hole.

## 4. The experimental development: from the past to the present

The first successful 1.3 mm VLBI experiment measuring Sgr A* was a single baseline observation between the IRAM telescopes on Pico Veleta (Spain) and Plateau de Bure (France) led by the MPIfR in Bonn [33][5]. The idea of imaging the event horizon in Sgr A* was first more widely discussed – though somewhat skeptically – at the Galactic Center conference in Tucson 1998 [e.g., see transcript of VLBI discussions in 61], followed by more discussions at other meetings. In 2004, we organized an informal evening session at the 30th anniversary meeting on the occasion of the discovery of Sgr A*[6], where the community was finally ready to embrace and support such an experiment[7]. The three speakers at the meeting, G. Bower, S. Doeleman and myself, started a series of telecons, further developing the idea of a global physics-like collaboration for black hole imaging. MIT Haystack observatory kept improving digital VLBI hardware for 3 and 1mm observing and proceeded further in a small ad hoc collaboration [12]. A little earlier, regular 3mm observing had already been made possible via a joint memorandum of understanding between multiple observatories, forming the Global mm-VLBI Array (GMVA, 2003), coordinated by the MPIfR Bonn.

Significant progress was made at lower frequencies. Bower et al. [5] were able to measure the intrinsic size at 7mm and 13 mm, showing for the first time that indeed Sgr A* decreases in size towards higher frequencies as predicted and that it would approach event horizon scales if extrapolated. This trend was soon confirmed by better VLBI measurements at 3mm [56].

[5] The observation was at low signal-to-noise and the size seemed much larger than the event horizon, hence, did not get too much attention. However, at the time the black hole mass was thought to be much smaller than it is now and the size expressed in Schwarzschild radii seemed larger than the shadow size. In fact, the measured size in $\mu$as at 230 GHz is consistent with current measurements within the error bars.

[6] `http://www.aoc.nrao.edu/~gcnews/GCconfs/SgrAstar30` — see small print after "Poster Contributions".

[7] In fact, expressed in the form of a vote and a signed declaration.

Doeleman et al. [12] published the first three-baseline experiment, demonstrating a source size of $37^{+16}_{-10}$ $\mu$as, i.e. smaller than the shadow size. Within the errors the results were consistent with the Krichbaum et al. [33] results, but more robust and with significantly smaller error bars. The triangle SMTO (Arizona) – CARMA (California) – JCMT/SMA (Hawaii) had a number of productive observing runs, showing asymmetric structure and evidence for homogenous magnetic fields [24, 31]. Also, a four-baseline experiment, involving the APEX telescope, succeeded, revealing evidence for more event-horizon scale structure [34, Lu et al. 2018, in prep.].

Community white papers to support mm-VLBI efforts were published in the US, Europe, and Asia [17, 23, 58, 1] and mm-VLBI was included in the European Science Vision for astronomy[8] in 2007 and in the US Astro2010 decadal review. In the summer of 2017 a formal memorandum of understanding was signed by 13 stakeholder institutions with about 150 individual scientists to form the Event Horizon Telescope Consortium (EHTC), which currently is conducting observations with 8 telescopes (IRAM 30m, JCMT, SMA, SMTO, SPT, LMT, ALMA, APEX) on six mountains[9]. Besides outfitting telescopes with proper VLBI-equipment, getting ALMA VLBI-ready had been another important step [40]. The EHTC experiments are ongoing and one has to await further data analysis, before any results can be discussed.

Despite the high bandwidth and some large telescopes, the limited number of simultaneous baselines will make imaging still difficult. Simulations by Lu et al. [35] suggest that images made during a single day suffer significant uncertainties from source variability. Substructures and variations in the scattering screen will amplify this uncertatiny further. Hence, multiple observing epochs may need to be added for a good image.

In the end, one will have to fit astrophysical shadow images to the experimental data [e.g., 8] to derive black hole parameters and to verify GR. However, the variations in those models lead to an unknown systematic error that still needs to be quantified better. Assuming GR is correct, shadow imaging could lead to an accuracy of up to 10 % [53] in confirming the null-hypothesis that a GR event horizon exists. If, however, GR itself is to be tested against alternative models, it may not be that easy to discriminate between GR shadow images and non-GR images [43].

Hence, like in cosmology, where multiple experiments are combined to provide, e.g., constraints on the nature of dark energy, it may also become important to add complementary information to the VLBI data. Psaltis, Kramer, & Wex (2016) show that measuring a pulsar in the Galactic center would be ideal for this. Indeed, after decades of searches, a first pulsar has been found [13] but closer and older ones are needed to measure the spacetime around Sgr A* accurately. The SKA, but also ALMA, could detect more pulsars in the future. Moreover, more and better stellar orbits will also constrain spacetime in a significant way. Here ESO's new near-infrared interferometer, GRAVITY, will make a big impact [29].

## 5. The future: Africa mm-wave telescope and space VLBI

The global millimeter VLBI array will further expand. ASIAA plans to add the Greenland telescope (GLT) for observations of M87 [39] and the BlackHoleCam team [28] together with IRAM, is preparing to phase up[10] the NOEMA array (Plateau de Bure, France) for VLBI. Also, the University of Arizona Kitt Peak telescope and the LLAMA dish (Brazil, Argentina) may be added soon.

Moreover, we are currently investigating possibilities to add an African mm-wave telescope (AMT) to the array. The Gamsberg mountain in Namibia is a potential site [2]. First simulations [55, see also Fig. 2] indicate that an Africa telescope would add more robustness and quality to

[8] `http://www.eso.org/public/news/eso0744/`
[9] The term "Event Horizon Telescope" came out of a coffee table discussion at the AAS meeting in Long Beach in 2009.
[10] Phasing-up an array means to add the signals from individual antennas in an array in phase, such that the telescopes act collectively as a large single dish.

IOP Conf. Series: Journal of Physics: Conf. Series **942** (2017) 012001 doi:10.1088/1742-6596/942/1/012001

the imaging. If multiple epochs are added, the AMT – despite being smaller – would become as important as ALMA for the image quality. In the case of long observations and high signal-to-noise ratio, the distribution of the telescopes is more important than the sensitivity of individual telescopes. Of course, more telescopes in Africa, e.g., near Kilimanjaro in Tanzania, or on the Drakensberg in Lesotho, would improve the quality even further.

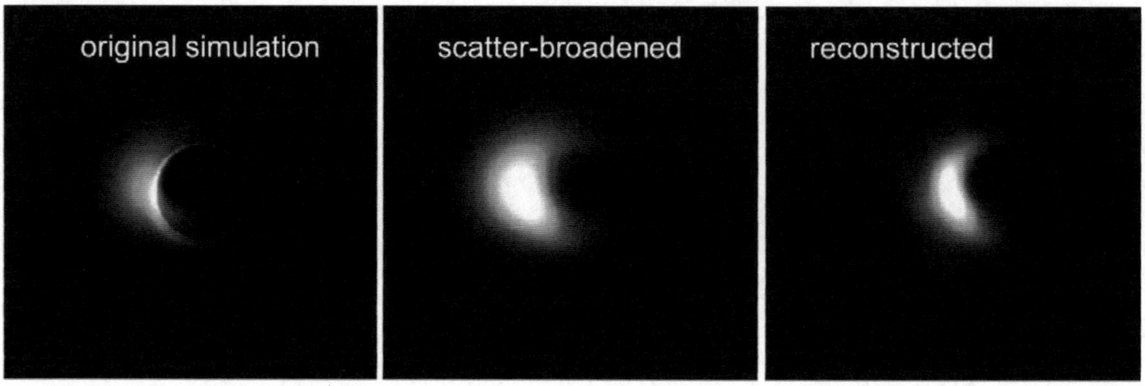


Figure 2 Left: Average image of the time variable black hole "movie" used in Lu et al. (2016), b) same image convolved with scattering kernel to account for interstellar scattering, c) synthetic image recovered with an array including SMA, SMT, LMT, ALMA, IRAM 30m, NOEMA, SPT, and an Africa telescope (AMT) using the procedure described in Lu et al. and averaging 8 epochs.

Of course, as always, space is the final frontier. Roelofs et al. [55] performed simulations using a two-element space-to-space interferometer, the Event Horizon Imager (EHI), consisting of two spacecraft at about 15,000 km. The telescopes are separated by some 50 km in altitude (Fig. 3) and observe at 690 GHz. At these frequencies interstellar scattering is almost negligible ($\sim 2\mu$as), the source will be very optically thin, and a ground-based VLBI array is unfeasible due to strong and variable tropospheric absorption. Having only a single space-to-space baseline, rather than a fleet of spacecraft, greatly reduces the complexity.

While orbiting the earth, the spacecraft slowly drift apart due to different orbital speeds. Eventually, the connecting baseline will have assumed all orientations and separations, giving one a spiral pattern in the uv-plane — the Fourier plane of the image. A filled uv-plane greatly improves image reconstruction in interferometric measurements. For realistic receiver noise and source fluxes, two 6 m dishes could obtain a very detailed picture of the black hole shadow (Fig. 3) with 4 $\mu$as resolution, when integrating over several months or even years — thereby also integrating over source variability. The constant features are those produced by GR.

## 6. Conclusion

All theoretical and experimental data support the idea that imaging the shadow of the black hole in the Galactic Center (and probably also in M87) should be possible with global radio interferometers at 230 GHz and higher. For testing general relativity, it is crucial that the emission region indeed extends well inside the ISCO, otherwise we would mainly see a hole in the accretion flow, rather than a black hole shadow. Hence, understanding the astrophysical model remains important. Current GRMHD models with a radiatively inefficient accretion flow and a jet outflow already explain the data for Sgr A* and M87* well. The newly formed Event Horizon Telescope Consortium is now attempting to make the first such images. Simulations suggest that multiple epochs may be necessary to obtain a robust image. If successful they could verify the prediction of an event horizon to within 10% accuracy, assuming GR is correct.

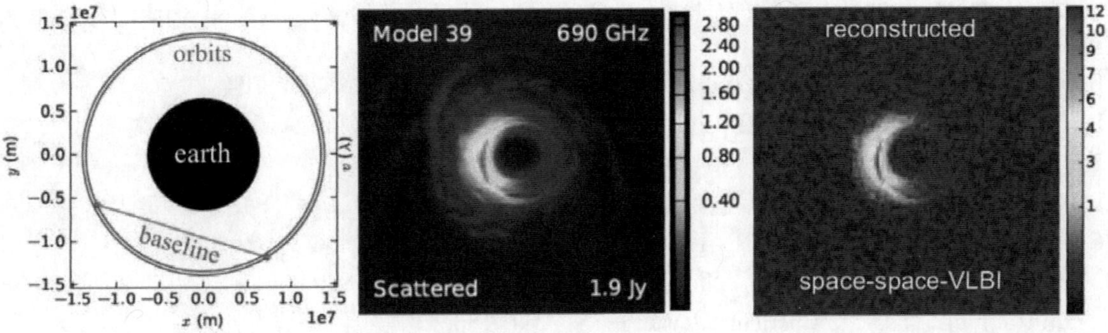


Figure 3 Simulated Space VLBI (EHI) image from Roelofs et al. [55]. Left: Schematic of the orbits (blue) of the two spacecraft (red points), Middle: Model image at 690 GHz from Moscibrodzka et al. [46], Right: images recovered from synthetic image data integrated over 2 years. The synthetic data were generated using the EHT imaging software from Chael et al. [10].

Testing alternative theories of gravity would benefit from additional inputs, such as better stellar orbits and ideally, a pulsar around the black hole. Images can be improved by adding additional telescopes to the global array. An African mm-wave telescope (AMT) would add crucial new baselines, producing better and more robust images. Ultimately, a space interferometer, such as the Event Horizon Imager operating at THz frequencies, could produce a quantum leap in black hole imaging. This may be possible even with just two spacecraft. Hence, it is probably more a question of when, rather than whether, the first good images of black holes will arrive. Since most supermassive black holes stay around for a long time, those images can and will become better with time. This provides us with the golden opportunity to conduct more and more detailed tests of gravity in its most extreme limit.

**Acknowledgments**
I am grateful for comments from F. Roelofs, M. Moscibrodzka, A. Roy, N. Nagar, G. Crew, and other colleagues. The author is supported by the European Research Council (ERC) Synergy Grant BlackHoleCam (Grant 610058), which is jointly led by the author togehther with L. Rezzolla (Univ. Frankfurt), and M. Kramer (MPIfR Bonn).

## References
[1] Asada, K., Kino, M., Honma, M., et al. 2017, ArXiv e-prints, 1705.04776

[2] Backes, M., Müller, C., Conway, J. E., et al. 2016, in Proceedings of the 4th Annual Conference on High Energy Astrophysics in Southern Africa (HEASA 2016). January 13th, 2016. South African Astronomical Observatory (SAAO), Cape Town, South Africa., 29

[3] Balick, B. & Brown, R. L. 1974, ApJ, 194, 265

[4] Bardeen, J. M. 1973, in Black Holes, ed. C. DeWitt & B. S. DeWitt (New York: Gordon & Breach), 215–239

[5] Bower, G. C., Falcke, H., Herrnstein, R. M., et al. 2004, Science, 304, 704

[6] Bower, G. C., Wright, M. C. H., Falcke, H., & Backer, D. C. 2003, ApJ, 588, 331

[7] Bridle, A. H. & Perley, R. A. 1984, ARA&A, 22, 319

[8] Broderick, A. E., Johannsen, T., Loeb, A., & Psaltis, D. 2014, ApJ, 784, 7

[9] Campbell, G. A. & Matzner, R. A. 1973, Journal of Mathematical Physics, 14, 1

[10] Chael, A. A., Johnson, M. D., Narayan, R., et al. 2016, ApJ, 829, 11

[11] Cunningham, C. T. & Bardeen, J. M. 1973, ApJ, 183, 237

[12] Doeleman, S. S., Weintroub, J., Rogers, A. E. E., et al. 2008, Nature, 455, 78

[13] Eatough, R. P., H. Falcke, Karuppusamy, R., & et al. 2013, Nature, doi:10.1038/nature12499

[14] Eckart, A. & Genzel, R. 1996, Nature, 383, 415

[15] Falcke, H. 1999, in ASP Conf. Ser. 186: The Central Parsecs of the Galaxy, ed. H. Falcke, A. Cotera, W. Duschl, F. Melia, & M. J. Rieke (San Francisco: Astronomical Society of the Pacific), 148

[16] Falcke, H., Goss, W. M., Matsuo, H., et al. 1998, ApJ, 499, 731

[17] Falcke, H., Laing, R., Testi, L., & Zensus, A. 2012, The Messenger, 149, 50

[18] Falcke, H., Mannheim, K., & Biermann, P. L. 1993, A&A, 278, L1

[19] Falcke, H. & Markoff, S. 2000, A&A, 362, 113

[20] Falcke, H. & Markoff, S. B. 2013, Classical and Quantum Gravity, 30, 244003

[21] Falcke, H., Melia, F., & Agol, E. 2000, ApJ, 528, L13

[22] Ferrari, A. 1998, ARA&A, 36, 539

[23] Fish, V., Alef, W., Anderson, J., et al. 2013, ArXiv e-prints, 1309.3519

[24] Fish, V. L., Johnson, M. D., Doeleman, S. S., et al. 2016, ApJ, 820, 90

[25] Genzel, R., Eisenhauer, F., & Gillessen, S. 2010, Reviews of Modern Physics, 82, 3121

[26] Ghez, A. M., Klein, B. L., Morris, M., & Becklin, E. E. 1998, ApJ, 509, 678

[27] Gillessen, S., Plewa, P. M., Eisenhauer, F., et al. 2017, ApJ, 837, 30

[28] Goddi, C., Falcke, H., Kramer, M., et al. 2017, International Journal of Modern Physics D, 26, 1730001

[29] Gravity Collaboration, Abuter, R., Accardo, M., et al. 2017, A&A, 602, A94

[30] Johannsen, T. 2016, Classical and Quantum Gravity, 33, 124001

[31] Johnson, M. D., Fish, V. L., Doeleman, S. S., et al. 2015, Science, 350, 1242

[32] Konoplya, R., Rezzolla, L., & Zhidenko, A. 2016, Phys. Rev. D, 93, 064015

[33] Krichbaum, T. P., Graham, D. A., Witzel, A., et al. 1998, A&A, 335, L106

[34] Krichbaum, T. P., Roy, A., Lu, R.-S., et al. 2014, in Proceedings of the 12th European VLBI Network Symposium and Users Meeting (EVN 2014). 7-10 October 2014. Cagliari, Italy., 13

[35] Lu, R.-S., Roelofs, F., Fish, V. L., et al. 2016, ApJ, 817, 173

[36] Luminet, J.-P. 1979, A&A, 75, 228

[37] Lynden-Bell, D. & Rees, M. J. 1971, MNRAS, 152, 461

[38] Marrone, D. P., Moran, J. M., Zhao, J.-H., & Rao, R. 2007, ApJ, 654, L57

[39] Matsushita, S., Asada, K., Blundell, R., et al. 2015, IAU General Assembly, 22, 2251138

[40] Matthews, L. D., Crew, G. B., Doeleman, S. S., et al. 2017, ArXiv e-prints

[41] Melia, F. & Falcke, H. 2001, ARA&A, 39, 309

[42] Mezger, P. G., Zylka, R., Salter, C. J., et al. 1989, A&A, 209, 337

[43] Mizuno, Y., Younsi, Z., Fromm, C. M., et al. 2018, tbd., to be submitted

[44] Moscibrodzka, M. 2017, in IAU Symposium, Vol. 322, The Multi-Messenger Astrophysics of the Galactic Centre, ed. R. M. Crocker, S. N. Longmore, & G. V. Bicknell, 43–49

[45] Moscibrodzka, M. & Falcke, H. 2013, A&A, 559, L3

[46] Moscibrodzka, M., Falcke, H., Shiokawa, H., & Gammie, C. F. 2014, A&A, 570, A7

[47] Moscibrodzka, M., Gammie, C. F., Dolence, J. C., Shiokawa, H., & Leung, P. K. 2009, ApJ, 706, 497

[48] Narayan, R., Yi, I., & Mahadevan, R. 1995, Nature, 374, 623

[49] Novikov, I. D. & Thorne, K. S. 1973, in Black Holes (Les Astres Occlus), ed. C. Dewitt & B. S. Dewitt, 343–450

[50] Page, D. N. & Thorne, K. S. 1974, ApJ, 191, 499

[51] Porth, O., Olivares, H., Mizuno, Y., et al. 2017, Computational Astrophysics and Cosmology, 4, 1

[52] Psaltis, D., Wex, N., & Kramer, M. 2016, ApJ, 818, 121

[53] Psaltis, D., zel, F., Chan, C.-K., & Marrone, D. P. 2015, ApJ, 814, 115

[54] Reid, M. J. & Brunthaler, A. 2004, ApJ, 616, 872

[55] Roelofs, F., Falcke, H., Brinkerink, C., et al. 2017, A&A, submitted

[56] Shen, Z.-Q., Lo, K. Y., Liang, M.-C., Ho, P. T. P., & Zhao, J.-H. 2005, Nature, 438, 62

[57] Sulentic, J. W., Marziani, P., & D'Onofrio, M. 2012, in Astrophysics and Space Science Library, Vol. 386, Fifty Years of Quasars: From Early Observations and Ideas to Future Research, ed. M. D'Onofrio, P. Marziani, & J. W. Sulentic, 549

[58] Tilanus, R. P. J., Krichbaum, T. P., Zensus, J. A., et al. 2014, ArXiv e-prints, 1406.4650

[59] Viergutz, S. U. 1993, A&A, 272, 355

[60] Yuan, F., Markoff, S., & Falcke, H. 2002, A&A, 383, 854

[61] Zensus, J. A. & Falcke, H. 1999, in Astronomical Society of the Pacific Conference Series, Vol. 186, The Central Parsecs of the Galaxy, ed. H. Falcke, A. Cotera, W. J. Duschl, F. Melia, & M. J. Rieke (San Francisco: Astronomical Society of the Pacific), 118

IOP Conf. Series: Journal of Physics: Conf. Series **942** (2017) 012002 doi:10.1088/1742-6596/942/1/012002

# Accelerating Black Holes

**Ruth Gregory**

Centre for Particle Theory, Durham University, South Road, Durham, DH1 3LE, UK
Perimeter Institute, 31 Caroline Street North, Waterloo, ON, N2L 2Y5, Canada

E-mail: r.a.w.gregory@durham.ac.uk

**Abstract.** In this presentation, I review recent work [1, 2] with Mike Appels and David Kubizňák on thermodynamics of accelerating black holes. I start by reviewing the geometry of accelerating black holes, focussing on the conical deficit responsible for the 'force' causing the black hole to accelerate. Then I discuss black hole thermodynamics with conical deficits, showing how to include the tension of the deficit as a thermodynamic variable, and introducing a canonically conjugate *thermodynamic length.* Finally, I describe the thermodynamics of the slowly accelerating black hole in anti-de Sitter spacetime.

## 1. Introduction

Black holes have proved to be perennially fascinating objects to study, both from the theoretical and practical points of view. We have families of exact solutions in General Relativity and beyond, and an ever better understanding of their phenomenology via numerical and observational investigations. However, in all of these situations the description of the black hole is as an isolated object, barely influenced by its sometimes extreme environment, its only possible response being to grow by accretion. The familiar exact solutions in GR describe precisely this – an isolated, topologically spherical, solution to the Einstein equations. The black hole is the ultimate slippery object – to accelerate, we must be able to 'push' or 'pull' on the actual event horizon, yet, by its very nature, anything touching the event horizon must be drawn in, unless it travels locally at the speed of light. Translating this to a physical object touching the event horizon, this means that the energy momentum tensor must locally have the form $T^0_0 \sim T^r_r$. Luckily, we have a candidate object that satisfies this requirement.

A cosmic string [3] is a very thin, quasi-linear object, with an energy momentum dominated by its mass per unit length, and a string tension equal in magnitude to this energy. Cosmic strings can be formed in field theories with non-simply connected vacua, and gravitationally produce no long range local curvature, but generate an overall global conical deficit in the spatial sections orthogonal to the string, thus a string, effectively, is a local conical deficit in the spacetime. It turns out that this conical deficit is precisely what accelerates the black hole.

## 2. Accelerating black holes

The accelerating black hole metric has been known for many years, though perhaps less widely outside the classical GR community, thus it is worth briefly reviewing its structure. It is described

IOP Conf. Series: Journal of Physics: Conf. Series **942** (2017) 012002 doi:10.1088/1742-6596/942/1/012002

by the *C-metric* [4, 5, 6], which can be written in Schwarzschild-like coordinates [7] as

$$ds^2 = \frac{1}{(1+Ar\cos\theta)^2}\left\{f(r)dt^2 - \frac{dr^2}{f(r)} - r^2\left[\frac{d\theta^2}{g(\theta)} + g(\theta)\sin^2\theta\frac{d\phi^2}{K^2}\right]\right\} \tag{1}$$

where $\phi$ has periodicity $2\pi$, and

$$f(r) = (1-A^2r^2)\Big(1-\frac{2m}{r}\Big) + \frac{r^2}{\ell^2}\,, \qquad g(\theta) = 1 + 2mA\cos\theta\,, \tag{2}$$

where $m$ represents the mass scale of the black hole, $A$ the acceleration, $K$ the conical deficit, and $\ell$ the (negative) cosmological constant, via $\Lambda = -3/\ell^2$. It is worth briefly deconstructing the C-metric to highlight the importance of the various parameters to the geometry and physics of the solution.

The function $f(r)$ determines the horizon structure of the black hole; whether there are one, two, or no horizons, depends on the mass parameter and the competition between the acceleration and cosmological constant. First, to understand the role of $K$, set $A \to 0$. Then (1) is clearly identifiable as the Schwarzschild-AdS solution, with the proviso that $\phi$ is now divided by $K$. Focussing on the axis through the black hole shows that that circles around this axis have a circumference $2\pi(1-1/K)$ times their proper radius, instead of the expected $2\pi$ of a locally flat space. Thus, the axis contains a conical deficit

$$\delta = 2\pi\left(1-\frac{1}{K}\right) = 8\pi\mu \tag{3}$$

running through the black hole [8] that can be replaced by a non-singular cosmic string [9].

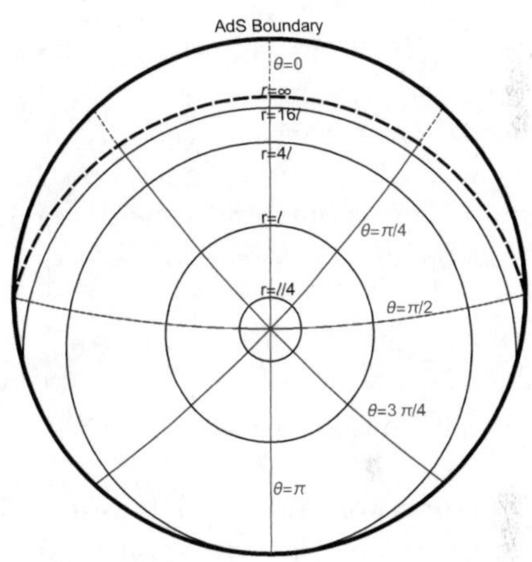


**Figure 1.** AdS spacetime in slowly accelerating coordinates.

Now consider the acceleration parameter, $A$. This in part counterbalances the effect of the cosmological constant in $f(r)$, and if $m$ is nonzero, gives a distortion of the two-spheres at constant $r$. To isolate the effect of $A$, setting $m = 0$ and $K = 1$ in (1) and keeping $A\ell < 1$ means that there is no acceleration horizon, and a simple coordinate transformation

$$1 + \frac{R^2}{\ell^2} = \frac{1+(1-A^2\ell^2)r^2/\ell^2}{(1-A^2\ell^2)\Omega^2} \qquad ; \qquad R\sin\Theta = \frac{r\sin\theta}{\Omega} \tag{4}$$

gives pure AdS spacetime [10]. The C-metric coordinates in (1) represent an off-centre perspective of AdS – they are the coordinates of an observer hovering at fixed distance from the centre of AdS, who is actually accelerating (see figure 1), but not sufficiently strongly to form an acceleration horizon.

Putting together, $A$ gives the acceleration of the black hole that counterbalances the effect of the cosmological constant in the Schwarzschild potential. The acceleration also gives an imbalance between the North and South axes that now have different conical deficits:

$$\begin{aligned} &\bullet\quad \theta \to \theta_+ = 0 \quad ds^2_{\theta,\phi} \propto d\theta^2 + \frac{(1+2mA)^2}{K^2}\theta^2 d\phi^2 \\ &\bullet\quad \theta \to \theta_- = \pi \quad ds^2_{\theta,\phi} \propto d\theta^2 + \frac{(1-2mA)^2}{K^2}(\pi-\theta)^2 d\phi^2 \end{aligned} \tag{5}$$

leading to the interpretation of the acceleration being induced by the imbalance of tension between the North and South axes:

$$8\pi\mu_\pm = \delta_\pm = 2\pi\left(1 - \frac{g(\theta_\pm)}{K}\right) = 1 - \frac{1}{K}\left(1 \pm 2mA + e^2A^2\right) \tag{6}$$

Conventionally, we choose to make the N axis regular by setting $K = 1 + 2mA$, leaving the deficit on the S axis corresponding to a cosmic string of tension $\mu_S = mA/K$ [11] (see figure 2. We will also demand that we have a *slowly accelerating* black hole, i.e. the acceleration is sufficiently gentle that we do not get an acceleration horizon (almost equivalent to $A\ell < 1$) [12]. The geometry therefore looks like a black hole in AdS displaced from the centre, the force required to hold it being supplied by the cosmic string or conical deficit running from the horizon to the boundary. Interpreting this tension as the force, and $A$ as the acceleration, Newton's Law suggests that the mass[1] of the black hole should then be identified as $M = m/K$, so that $\mu = F = MA$ – we will see presently that this is exactly what the first law of thermodynamics requires.

**Figure 2.** A representation of the accelerating black hole.

## 3. Thermodynamics with conical defects

Before discussing the accelerating black hole, set $A = 0$ and look at the effect on the Schwarzschild-AdS black hole of the conical deficit (see [14] for early work with fixed tension deificits). The horizon is defined as $f(r) = 0$, so we examine incremental changes in the black hole

$$f(r_+ + \delta r_+) = f'(r_+)\delta r_+ + \frac{\partial f}{\partial m}\delta m + \frac{\partial f}{\partial \ell}\delta\ell = 0 \tag{7}$$

where we are allowing not only for a change in the black hole mass, but a possible change in the cosmological constant, as might occur for example when vacuum energy is due to a scalar field that is allowed to vary (see [15] for a discussion of black hole thermodynamics in this situation).

Defining temperature in the usual way via surface gravity, $T = f'(r_+)/4\pi$, we see that the first term in (7) will be related to a change in entropy, however, now the entropy, given by the area of the horizon [16], also depends on $K$:

$$S = \frac{\pi r_+^2}{K} \to f'(r_+)\delta r_+ = \frac{2K}{r_+}\left(T\delta S + \frac{r_+^2 f'(r_+)}{4}\frac{\delta K}{K^2}\right) \tag{8}$$

To identify the physics in changing $K$, note that tension was related to $K$, (3), from which we deduce

$$\delta\mu = \frac{\delta K}{4K^2} \tag{9}$$

[1] Note that unlike the rapidly accelerating black hole, this is a genuine ADM-style mass, and not a "re-arrangement of dipoles" as discussed in [13] where a boost mass was introduced.

IOP Conf. Series: Journal of Physics: Conf. Series **942** (2017) 012002 doi:10.1088/1742-6596/942/1/012002

Finally, writing the thermodynamic pressure [17, 18, 19, 20]

$$P = -\Lambda = \frac{3}{8\pi\ell^2} \qquad V = \frac{4\pi r_+^3}{3K} \tag{10}$$

allows us to identify a first law from (7)

$$\delta(\frac{m}{K}) = T\delta S + 2(m - r_+)\delta\mu + V\delta P \tag{11}$$

Leading to the definition of mass for the black hole $M = m/K$, as suggested by Newton's Law. Indeed, taking this definition of $M$ also gives a Smarr formula [21] for the black hole:

$$M = 2TS - 2PV \tag{12}$$

Let us look more closely at this first law. The term multiplying the variation in tension has the form of a length, and we define this to be the *thermodynamic length*[2] associated to the tension charge of the string:

$$\lambda = r_+ - m \tag{13}$$

Note, we do not include the factor of 2, but instead have the length of the string from each pole – this will be particularly relevant when we look at the accelerating black hole that has different deficits at each pole. This reinforces the interpretation of $M$ as the enthalpy [17] of the black hole: if the black hole grows, it swallows some string, but has also displaced the same amount of energy from the environment.

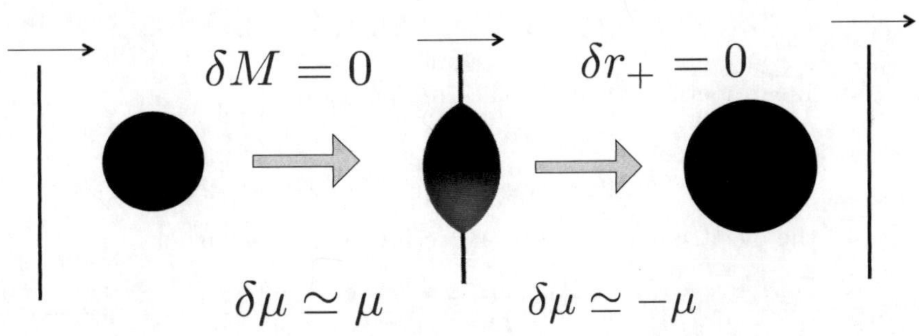


**Figure 3.** A black hole capturing a piece of cosmic string.

Does this definition and the first law make sense? In order to explore this, consider the situation of a cosmic string interacting with a black hole, first proposed in [23], and for simplicity let $\Lambda = 0$, for which the first law reduces to

$$\delta M = T\delta S - 2\lambda\delta\mu \tag{14}$$

Initially, the string is outside the black hole, but moving towards it (see figure 3). The black hole then captures the string and thus acquires a positive conical deficit, however because the string is still moving, it eventually pulls off the black hole, leaving a purely spherical black hole behind. The first step of the process has a positive $\delta\mu$ and the second an equal but negative

[2] Note, a similar expression was obtained by Kastor and Traschen [22] in the context of an angular gravitational tension associated to the rotational symmetry of the black hole.

IOP Conf. Series: Journal of Physics: Conf. Series **942** (2017) 012002 doi:10.1088/1742-6596/942/1/012002

tension change, $-\delta\mu$. In the first stage, we would expect the total energy of the black hole to be constant, thus $\delta M = 0$, the physical intuition being that a cosmic string passing through a spherical shell of matter would not change the total energy throughout the process, For the black hole however, string cannot emerge from within the event horizon, and we have the interpretation of a segment of string being captured by the black hole, with the black hole increasing its mass accordingly.

In the first stage (taking $\mu \ll 1$), fixing $M$ implies

$$T\delta S = 2\lambda\mu \tag{15}$$

consistent with $\delta m = 4m\delta\mu$, and $S = 4\pi m^2/K$. Because the energy of the black hole has been fixed, the event horizon moves outwards to compensate for the conical deficit, resulting in an increase of entropy indicating this is an irreversible thermodynamic process. In the second stage, the string pulls off the black hole, so $\delta\mu = -\mu$, and since entropy cannot decrease, $M$ must increase:

$$\delta M = T\delta S + 2m\mu = \delta m + 4m\mu \tag{16}$$

in this case, it seems reasonable to suppose that the local geodesic congruence defining the event horizon cannot contract, giving $\delta m = 0$, and $\delta M = 4m\mu$. Thus, the net result of this adiabatic process is that the black hole has 'swallowed' the segment of cosmic string that fell inside its event horizon.

## 4. Thermodynamics of accelerating black holes

To find the first law for accelerating black holes, the idea is to follow the same steps: the horizon is defined by $f(r) = 0$, and consider small increments in horizon radius:

$$\delta f(r_+) = f'_+\delta r_+ - 2\frac{\delta m}{r_+}(1 - A^2r_+^2) - 2A\delta A r_+(r_+ - 2m) - 2\frac{r_+^2}{\ell^3}\delta\ell = 0 \tag{17}$$

Using the usual Euclidean method we can relate temperature to $f'_+$,

$$T = \frac{f'(r_+)}{4\pi} = \frac{1}{2\pi r_+^2}\left[m(1 - A^2r_+^2) + \frac{r_+^3}{\ell^2(1 - A^2r_+^2)}\right] \tag{18}$$

but now the area of the event horizon gives a more involved relation for $\delta r_+$:

$$\delta S = \delta\left(\frac{\pi r_+^2}{K(1 - A^2r_+^2)}\right) = \frac{2\pi r_+\delta r_+}{K(1 - A^2r_+^2)^2} + \frac{2\pi r_+^4 A\delta A}{K(1 - A^2r_+^2)^2} - \frac{\pi r_+^2}{(1 - A^2r_+^2)}\frac{\delta K}{K^2} \tag{19}$$

The expressions for the tensions (6) allow us to replace $\delta K/K^2 = 2(\delta\mu_+ + \delta\mu_-)$ and $m\delta A/K = -\delta\mu_+ + \delta\mu_- - A\delta M$, having identified $M = m/K$ as before. Putting together, this gives the First Law as (see also [24])

$$\delta M = V\delta P + T\delta S - \lambda_+\delta\mu_+ - \lambda_-\delta\mu_- \,. \tag{20}$$

where we now have a thermodynamic length associated to the piece of string attached at each pole:

$$\lambda_\pm = \frac{r_+}{1 \pm Ar_+} - KM \tag{21}$$

This obviously agrees with (13) for the string threading the black hole, where $r_+$ has now been replaced by $r_+/\Omega(r_+, \theta_\pm)$ at each pole. We also have a Smarr relation,

$$M = 2TS - 2PV\,, \tag{22}$$

as before. The string does not contribute to the overall mass of the black hole as the string is gravitationally 'neutral' – it has no overall Riemann curvature, the spacetime is locally asymptotically flat – the effect of the string being simply a global deficit angle.

IOP Conf. Series: Journal of Physics: Conf. Series **942** (2017) 012002 doi:10.1088/1742-6596/942/1/012002

## 5. Exploring thermodynamics

Having derived a sensible first law for the accelerating black hole, now let us briefly explore some consequences of the expressions (for a full discussion see [1, 2]). First consider the temperature of the accelerating black hole (18). Typically, we associate a temperature to the surface gravity of an horizon; apart from the black hole, horizons arise in the context of cosmology with a positive cosmological constant, and acceleration. Here we have explicitly taken our acceleration sufficiently low that we never get an acceleration horizon, but nonetheless, the effect of acceleration counterbalances the cosmological constant in the gravitational potential: $1/\ell^2 \to 1/\ell^2 - A^2 = 1/\ell_{\text{eff}}^2$. We could perhaps expect that the temperature of the black hole is somehow a combination of the inherent Hawking temperature and the Rindler temperature, but in this case we would expect a systematic shift with $A$, and this is not what happens. Instead, for smaller mass black holes the temperature decreases with acceleration, but for larger mass black holes it increases (see figure 4). To some extent this can be understood as a result of the varying tensions that are required to accelerate a fixed mass black hole differently.

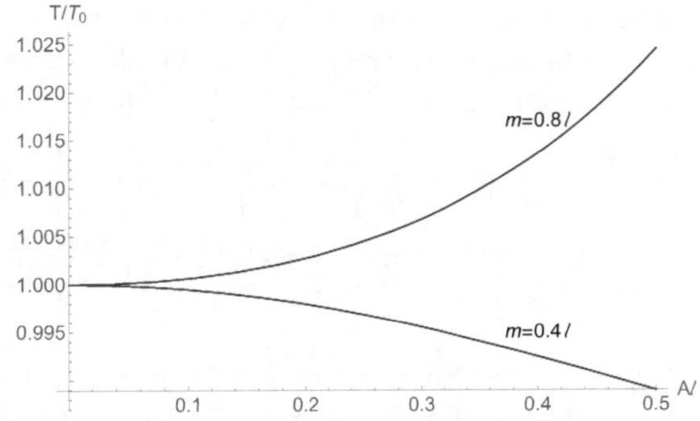


**Figure 4.** The variation of temperature of a fixed mass black hole with acceleration.

Another phenomenon we can explore is that of the Hawking-Page phase transition [25]. For small black holes in AdS, the vacuum curvature is subdominant to the local tidal curvature of the black hole, thus thermodynamics is similar to vacuum with black holes being thermodynamically unstable. For black holes larger than the AdS radius however, the local tidal curvature becomes subdominant to the vacuum curvature, and the black hole has positive specific heat. Thus in AdS, there is a minimum temperature for the black hole, and below this temperature only a radiation bath can be a solution to the Einstein equations at finite $T$. At higher temperatures, one can have either a small or large black hole (or radiation bath). The Hawking-Page transition occurs at $T > 1/\pi\ell$, where the large black hole is thermodynamically preferred to the radiation bath, as can be seen by plotting the Gibbs free energy as a function of temperature. For the

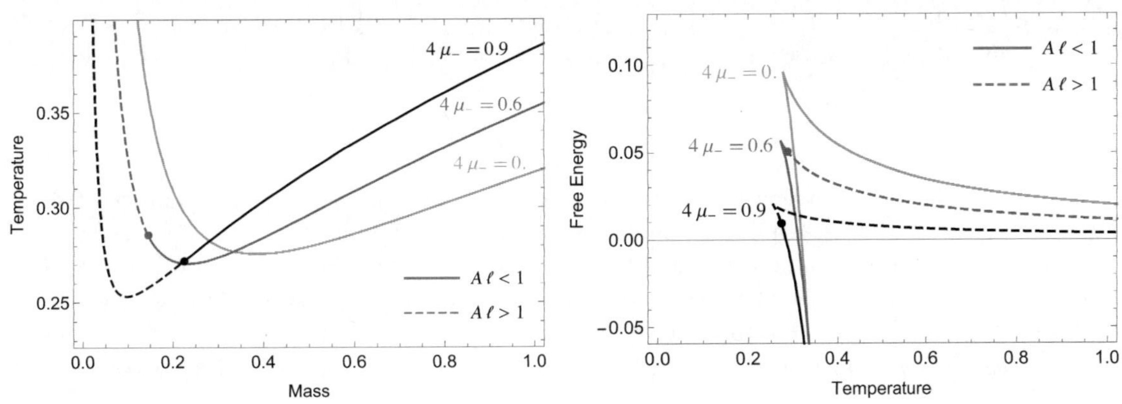


**Figure 5.** A plot of temperature as a function of mass (left), and the Gibbs free energy as a function of temperature (right).

accelerating black hole, we fix the tension of the string, and plot the temperature as a function of (thermodynamic) mass in figure 5. This plot explicitly demonstrates that acceleration makes a black hole of given mass *more* thermodynamically stable in the sense of positive specific heat. A plot of the corresponding Gibbs free energy is also shown.

For simplicity, I focussed primarily in this presentation on uncharged non-rotating black holes to demonstrate the procedure for determining the thermodynamic charges and potentials. One can add charge to the black hole, which now allows for a much richer thermodynamic phase structure. Briefly, the charges and potentials in the presence of electric charge become [2]):

$$\begin{aligned} M &= \frac{m}{K(1+e^2A^2)} \qquad Q = \frac{e}{K} \\ \Phi &= \frac{e}{r_+} - \frac{meA^2}{1+e^2A^2} \qquad \lambda_\pm = \frac{r_+}{1 \pm Ar_+} - \frac{m(1-e^2A^2)}{(1+a^2A^2)^2} \mp \frac{e^2A}{1+e^2A^2} \end{aligned} \tag{23}$$

with the entropy, thermodynamic pressure and volume unchanged in form.

With charge, there is a lower bound on the black hole mass from the extremal limit, thus at fixed $Q$, black holes have positive specific heat near this extremal limit, and at large mass, but in between can have negative specific heat if the charge is sufficiently low that the event horizon lies below the AdS scale. As with the uncharged black hole, acceleration makes the black hole more thermodynamically stable, as seen in figure 6.

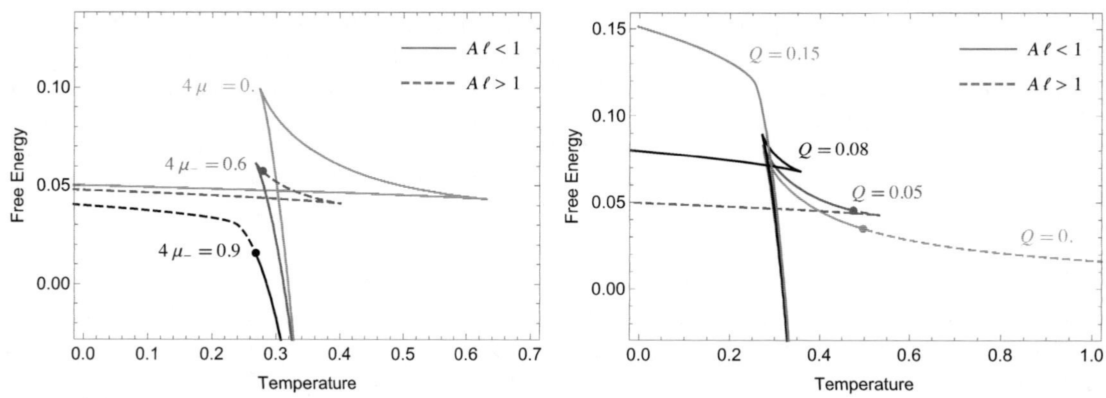


**Figure 6.** A plot of the free energy as a function of temperature for varying tension with $Q = 0.05\ell$ on the left, and varying charge with $4\mu_- = 0.3$ on the right.

## 6. Conclusion

To sum up: we have shown how to allow for varying tension in the thermodynamics of black holes. In accelerating black holes, if one does not allow the tension of the deficits to vary, then the various thermodynamic charges are not free to vary independently, and there is no guarantee that any thus derived thermodynamic quantities will be the correct ones, it is therefore mandatory to consider $\mu$ as a variable in order to correctly explore thermodynamics. Work on the thermodynamics of rotating accelerating black holes is reported on in the poster session [26].

## Acknowledgments

I would like to thank David Kubizňák and Mike Appels for collaboration on this project. This work was supported by STFC (Consolidated Grant ST/L000407/1), and also by the Perimeter Institute for Theoretical Physics. Research at Perimeter Institute is supported by

IOP Conf. Series: Journal of Physics: Conf. Series **942** (2017) 012002 doi:10.1088/1742-6596/942/1/012002

the Government of Canada through the Department of Innovation, Science and Economic Development Canada and by the Province of Ontario through the Ministry of Research, Innovation and Science.

## References

[1] Appels M, Gregory R, and Kubizňák D, Thermodynamics of Accelerating Black Holes, Phys. Rev. Lett. **117**, 131303 (2016) (*Preprint arXiv:1604.08812 [hep-th]*)
[2] Appels M, Gregory R, and Kubizňák D, Black Hole Thermodynamics with Conical Defects, JHEP **1705**, 116 (2017) (*Preprint arXiv:1702.00490 [hep-th]*).
[3] Vilenkin A, Cosmic strings and domain walls, Phys. Rep. **121** (1985) 263–315.
[4] Kinnersley W and Walker M, Uniformly accelerating charged mass in general relativity, Phys. Rev. D **2**, 1359 (1970).
[5] Plebanski J F and Demianski M, Rotating, charged, and uniformly accelerating mass in general relativity, Annals Phys. **98**, 98 (1976).
[6] Griffiths J B and Podolsky J, A New look at the Plebanski-Demianski family of solutions, Int. J. Mod. Phys. D **15**, 335 (2006) (*Preprint gr-qc/0511091*).
[7] Hong K and Teo E, A New form of the C metric, Class. Quant. Grav. **20**, 3269 (2003) (*Preprint gr-qc/0305089*).
[8] Aryal M, Ford L and Vilenkin A, Cosmic strings and black holes, Phys. Rev. **D34** (1986) 2263.
[9] Achucarro A, Gregory R, and Kuijken K, Abelian Higgs hair for black holes, Phys. Rev. **D52** (1995) 5729–5742, (*Preprint gr-qc/9505039*).
[10] Podolsky J, Accelerating black holes in anti-de Sitter universe, Czech. J. Phys. **52**, 1 (2002) (*Preprint gr-qc/0202033*).
[11] Gregory R and Hindmarsh M, Smooth metrics for snapping strings, Phys. Rev. **D52** (1995) 5598–5605, (*Preprint gr-qc/9506054*).
[12] Krtous P, Accelerated black holes in an anti-de Sitter universe, Phys. Rev. D **72**, 124019 (2005) (*Preprint gr-qc/0510101*).
[13] Dutta K, Ray S and Traschen J, Boost mass and the mechanics of accelerated black holes, Class. Quant. Grav. **23**, 335 (2006) (*Preprint hep-th/0508041*).
[14] E. A. Martinez and J. W. York, Jr., Thermodynamics of black holes and cosmic strings, Phys. Rev. D **42**, 3580 (1990).
[15] Gregory R, Kastor D and Traschen J, Black Hole Thermodynamics with Dynamical Lambda, *Preprint arXiv:1707.06586 [hep-th]*.
[16] Herdeiro C, Kleihaus B, Kunz J and Radu E, On the Bekenstein-Hawking area law for black objects with conical singularities, Phys. Rev. D **81**, 064013 (2010) (*Preprint arXiv:0912.3386 [gr-qc]*).
[17] Kastor D, Ray S and Traschen J, Enthalpy and the Mechanics of AdS Black Holes, Class. Quant. Grav. **26**, 195011 (2009) (*Preprint arXiv:0904.2765 [hep-th]*).
[18] Dolan B P, The cosmological constant and the black hole equation of state, Class. Quant. Grav. **28**, 125020 (2011) (*Preprint arXiv:1008.5023 [gr-qc]*).
[19] Dolan B P, Pressure and volume in the first law of black hole thermodynamics, Class. Quant. Grav. **28**, 235017 (2011) (*Preprint arXiv:1106.6260 [gr-qc]*).
[20] Kubizňák D and Mann R B, P-V criticality of charged AdS black holes, JHEP **1207**, 033 (2012) (*Preprint arXiv:1205.0559 [hep-th]*).
[21] Smarr L, Mass formula for Kerr black holes, Phys. Rev. Lett. **30**, 71 (1973) Erratum: [Phys. Rev. Lett. **30**, 521 (1973)].
[22] Kastor D and Traschen J, The Angular Tension of Black Holes, Phys. Rev. D **86**, 081501 (2012) (*Preprint arXiv:1207.5415 [hep-th]*).
[23] Bonjour F, Emparan R and Gregory R, Vortices and extreme black holes: The Question of flux expulsion, Phys. Rev. D **59**, 084022 (1999) (*Preprint gr-qc/9810061*).
[24] Astorino M, Thermodynamics of Regular Accelerating Black Holes, (*Preprint arXiv:1612.04387 [gr-qc]*).
[25] Hawking S W and Page D N, Thermodynamics of Black Holes in anti-De Sitter Space, Commun. Math. Phys. **87**, 577 (1983).
[26] Appels M, The ultra-spinning limit of accelerated black holes, *Poster, 3rd Karl Schwarzschild Meeting*; & Appels M, Gregory R, and Kubizňák D, to appear.

IOP Conf. Series: Journal of Physics: Conf. Series **942** (2017) 012003 doi:10.1088/1742-6596/942/1/012003

# Einstein-Maxwell-Chern-Simons Black Holes

**J Kunz[1], J L Blázquez-Salcedo[1], F Navarro-Lérida[2], E Radu[3]**

[1] Institut für Physik, Universität Oldenburg, D-26111 Oldenburg, Germany

[2] Dept. de Física Atómica, Molecular y Nuclear, Ciencias Físicas, Universidad Complutense de Madrid, E-28040 Madrid, Spain

[3] Departamento de Física da Universidade de Aveiro and CIDMA, Campus de Santiago, 3810-183 Aveiro, Portugal

E-mail: jutta.kunz@uol.de, jose.blazquez.salcedo@uol.de, fnavarro@fis.ucm.es, eugen.radu@ua.pt

**Abstract.** Black holes in 5-dimensional Einstein-Maxwell-Chern-Simons (EMCS) theory and their intriguing properties are discussed. For the special case of the CS coupling constant $\lambda = \lambda_{SG}$, as obtained from supergravity, a closed form solution is known for the rotating black holes. Beyond this supergravity value, the EMCS black hole solutions can e.g. exhibit nonuniqueness and form sequences of radially excited solutions. In the presence of a negative cosmological constant the black holes can possess an extra-parameter corresponding to a magnetic flux in addition to the mass, electric charge and angular momenta. This latter family of black holes possesses also a solitonic limit. Finally, a new class of squashed EMCS black hole solutions is discussed.

## 1. Introduction

The properties of black holes in higher dimensions can differ in many respects from those known in four dimensions. Most prominently, asymptotically flat vacuum black holes may possess a non-spherical horizon topology in more than four dimensions. But also for black holes with a spherical horizon suprises arise, as soon as a U(1) gauge field is coupled, i.e., Einstein-Maxwell (EM) black holes are considered. In odd dimensions the presence of a Chern-Simons (CS) term allows for further intriguing properties of the resulting Einstein-Maxwell-Chern-Simons (EMCS) black holes. In the following first asymptotically flat black hole solutions of EMCS theory will be discussed. Then a negative cosmological constant will be included, making contact with the celebrated AdS/CFT correspondence, a central topic of the conference.

## 2. Asymptotically Flat EMCS Black Holes

In odd dimensions $D = 2n + 1$ the Einstein-Maxwell action may be supplemented by an '$\mathcal{A}\mathcal{F}^n$' Chern-Simons term. In 5 dimensions the EMCS action reads

$$S = \int \frac{1}{16\pi G_5}\Big\{\sqrt{-g}\,(R - \mathcal{F}_{mn}\mathcal{F}^{mn}) - \frac{2\lambda}{3\sqrt{3}}\varepsilon^{mnpqr}\mathcal{A}_m\mathcal{F}_{np}\mathcal{F}_{qr}\Big\}d^5x \qquad (1)$$

with Newton's constant $G_5$, curvature scalar $R$, gauge potential $\mathcal{A}_m$, field strength tensor $\mathcal{F}_{mn}$, and CS coupling constant $\lambda$. In EM theory $\lambda = 0$, while $\lambda = 1$ in the bosonic sector of minimal

IOP Conf. Series: Journal of Physics: Conf. Series **942** (2017) 012003 doi:10.1088/1742-6596/942/1/012003

$D = 5$ supergravity. Variation of the action leads to the Einstein equations ($8\pi G = 1$)

$$G_{mn} = 2\left(\mathcal{F}_{mr}\mathcal{F}^{r}{}_{n} - \frac{1}{4}\mathcal{F}_{rs}\mathcal{F}^{rs}\right), \tag{2}$$

which are unchanged w.r.t. the pure Maxwell case, and to the Maxwell-CS equations

$$\nabla_n \mathcal{F}^{mn} + \frac{\lambda}{2\sqrt{3}}\epsilon^{mnpqr}\mathcal{F}_{np}\mathcal{F}_{qr} = 0. \tag{3}$$

Clearly, the CS term breaks the charge symmetry $Q \to -Q$ present in Maxwell theory.

### 2.1. $D = 5$ *Einstein-Maxwell Theory*

The $D$-dimensional generalizations of the Kerr black holes are given by the Myers-Perry (MP) black holes, which possess $N = \lfloor \frac{D-1}{2} \rfloor$ independent angular momenta $J_i$, $i = 1, \ldots, N$ [1]. The inclusion of a Maxwell field, however, most of the time prevents the construction of black hole solutions in closed form, even if in odd dimensions the angular momenta are chosen to be of equal magnitude, i.e., when only cohomogeneity-1 solutions are sought. In five dimensions an appropriate ansatz for the metric and the gauge potential is given by

$$ds^2 = -F_0(r)dt^2 + F_1(r)dr^2 + \frac{1}{4}F_2(r)(\sigma_1^2 + \sigma_2^2) + \frac{1}{4}F_3(r)(\sigma_3 - 2\omega(r)dt)^2, \tag{4}$$

$$\mathcal{A} = a_0(r)dt + a_\varphi(r)\frac{1}{2}\sigma_3 \tag{5}$$

with $\sigma_1^2 + \sigma_2^2 = d\bar{\theta}^2 + \sin^2\bar{\theta}d\psi^2$, $\sigma_3 = d\phi + \cos\bar{\theta}d\psi$, and functions $F_0, \ldots, F_3$, $\omega$, $a_0$ and $a_\varphi$.

In this case (i) perturbative calculations in the charge or in the rotation parameter can be performed, (ii) the solutions can be obtained numerically, or (iii) near-horizon solutions can be found in the extremal limit. To obtain the latter, an adequate ansatz for the metric and the gauge potential in five dimensions is given by [2]

$$ds^2 = v_1\left(\frac{dr^2}{r^2} - r^2dt^2\right) + v_2\left[\sigma_1^2 + \sigma_2^2 + v_3(\sigma_3 - krdt)^2\right], \tag{6}$$

$$\mathcal{A} = q_1 r dt + q_2(\sigma_3 - krdt) \tag{7}$$

with constants $v_1$, $v_2$, $v_3$, $q_1$, $q_2$ and $k$.

An analogous ansatz holds in more than five dimensions. Solving the resulting algebraic system of equations for the constants leads to two distinct solutions. The first solution starts at the extremal MP black hole and satisfies $J = \sqrt{2(D-3)}A_{\rm H}$, thus its angular momentum is proportional to its horizon area. The second solution starts at the extremal Reissner-Nordström (RN) black hole and satisfies $J = (D-1)J_{\rm H}$, thus its angular momentum is proportional to its horizon angular momentum. As seen in Figs. 1,2 both branches intersect at a critical solution (black cross) and continue beyond. Interestingly, only the name-giving parts of the MP and RN solutions up to the intersection point are realized globally [3, 4]. The domain of existence of these EM black holes is seen in Fig. 3 and corresponds to the area enclosed by the $\lambda = 0$ curve.

### 2.2. $D = 5$ *minimal supergravity*

For the case $\lambda = 1$ representing $D = 5$ minimal supergravity the black hole solutions are known in closed form [5, 6, 7]. The domain of existence is also seen in Fig. 3 and delimited by the $\lambda = 1$ curve. The charge symmetry $Q \to -Q$ is clearly broken here. The vertical line represents the extremal BMPV black holes [5]. For these black holes, as the charge $Q$ is kept fixed and the angular momentum $J$ increases, the mass $M$ remains constant.

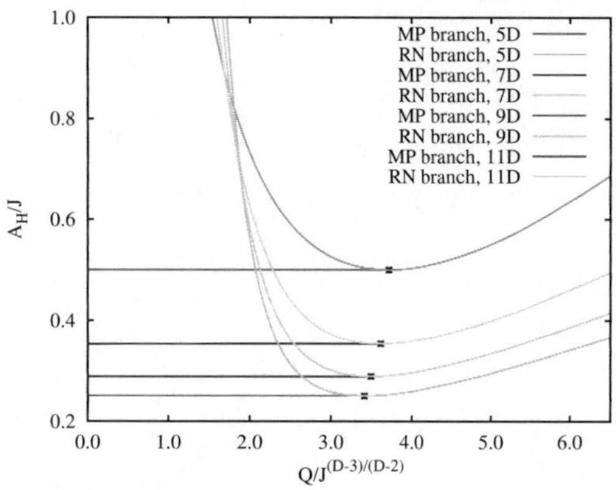


**Figure 1.** Odd-$D$ EM near-horizon solutions: area vs charge (scaled with appropriate powers of the angular momentum).

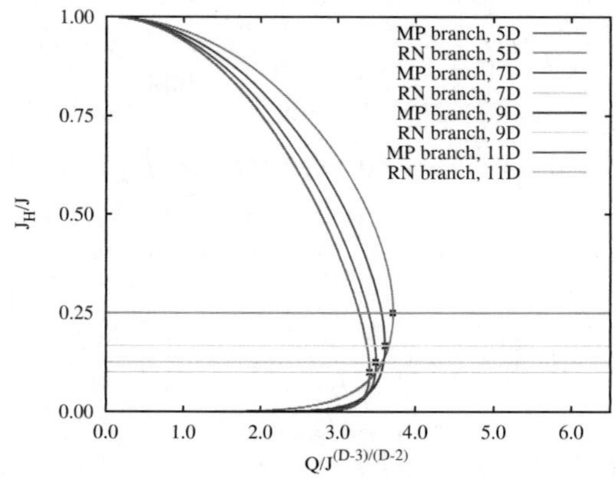


**Figure 2.** Same as Fig. 1 for the horizon angular momentum.

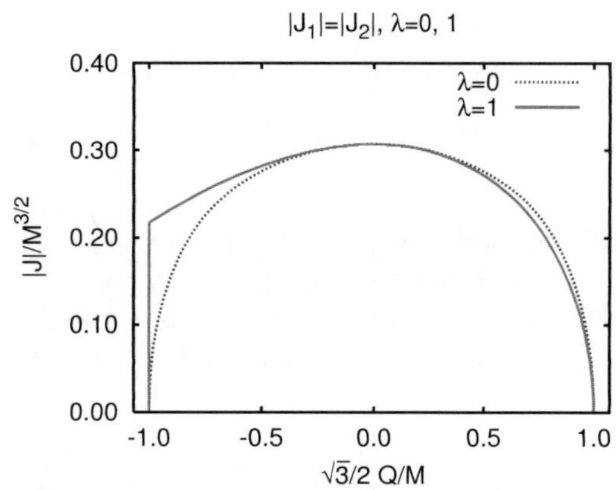


**Figure 3.** Domain of existence of $D = 5$ EMCS black holes: angular momentum vs charge (scaled with appropriate powers of the mass) for CS coupling $\lambda = 1$ and $\lambda = 0$.

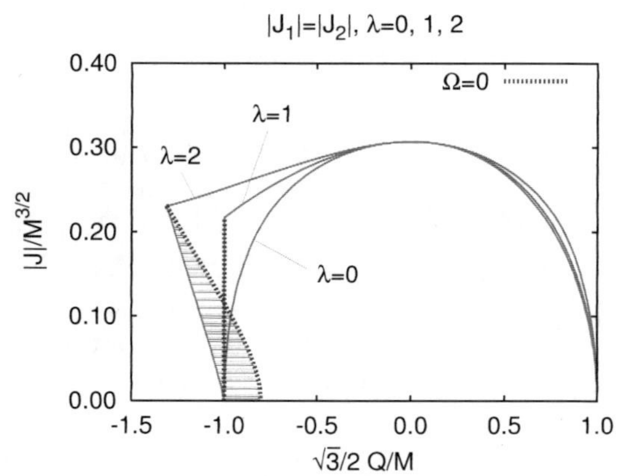


**Figure 4.** Same as Fig. 3 including CS coupling $\lambda = 2$. The shaded area represents counterrotating black holes. $\Omega = 0$ black holes are also indicated.

When considering the first law $dM = TdS + 2\Omega dJ + \Phi dQ$ for these BMPV black holes, one realizes that this implies that their horizon angular velocity $\Omega$ must be zero, while the global angular momentum is finite. Thus angular momentum is stored in the Maxwell field, where a negative fraction of the angular momentum resides behind the horizon. While the effect of rotation is to deform the horizon into a squashed 3-sphere [8]. The set of BMPV solutions ends at a critical solution with vanishing area.

### *2.3. $D = 5$ EMCS theory: $\lambda \neq 1$*

Let us now consider the CS coupling as a free parameter and increase it above the supergravity value [9]. Then it becomes clear that the supergravity value represents the borderline between stability and instability, since at $\lambda = 1$ a zero mode is present, leading to a rotational instability for larger values of $\lambda = 1$ [8, 9].

Then solutions with vanishing horizon angular velocity no longer form a part of the boundary

IOP Conf. Series: Journal of Physics: Conf. Series **942** (2017) 012003 doi:10.1088/1742-6596/942/1/012003

of the domain of existence (except for the critical solution with vanishing area at the cusp), but reside well within the domain of existence. Now this part of the boundary is formed by a new type of black hole solutions. These are counterrotating in the sense, that the horizon angular velocity and the global angular momentum carry opposite signs [9]. In Fig. 4 all counterrotating black holes are represented by the shaded area.

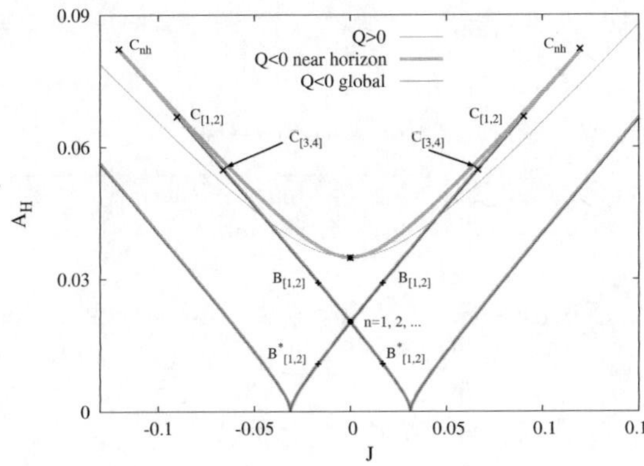


**Figure 5.** $D = 5$ EMCS near-horizon and global solutions: area vs angular momentum ($\lambda = 5$). The asterisk marks the extremal static RN black hole.

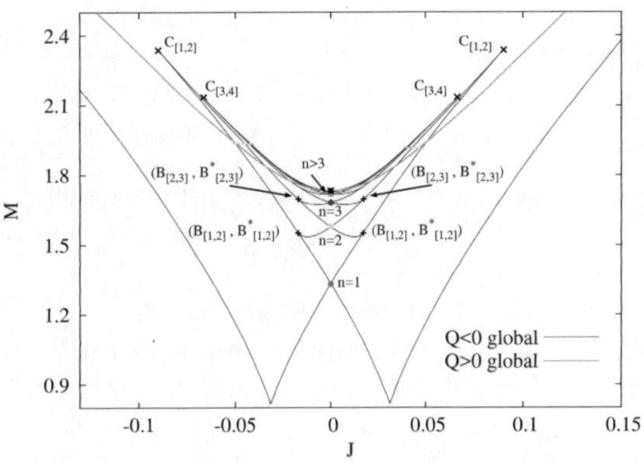


**Figure 6.** $D = 5$ EMCS global black hole solutions: mass vs angular momentum ($\lambda = 5$). Bifurcation points $B$ and cusps $C$ are marked.

For CS coupling $\lambda > 2$ further surprises appear. First of all, uniqueness of black holes with spherical horizon topology no longer holds [9]. Here distinct black holes with the same global charges are present. Second, there arises a strong mismatch between the near-horizon solutions and the global solutions [10, 11]. In fact, there are near-horizon solutions that correspond to (i) no global solution, (ii) one global solution, and (iii) many global solutions (possibly even infinitely many).

The area versus the angular momentum of the near-horizon solutions is illustrated in Fig. 5 for CS coupling $\lambda = 5$ for positive and negative charge. For comparison also the global solutions are included in the figure. Furthermore cusps and bifuraction points are noted. In particular, while well connected with the branches of near-horizon solutions, the RN solution is isolated from the $Q < 0$ global solutions. Moreover, the $Q < 0$ near-horizon solutions exhibit a non-static $J = 0$ solution, which corresponds to a set of distinct non-static global $J = 0$ solutions, that can be labeled by an integer $n$. Thus there are black holes with a rotating horizon but vanishing global angular momentum. Also, the $Q < 0$ solutions always contain a degenerate ($J \to -J$) zero area solution.

The mass versus the angular momentum of the corresponding global solutions is exhibited in Fig. 6. Here the degeneracy present in the area versus angular momentum plot is lifted. Instead an intricate branch structure of the global solutions is revealed. In fact, in a regular manner, new branches of solutions arise, which form cusps and then bifurcate with previous branches. The cusps $C$ and bifurcation points $B$ are counted and labeled by integers in the figure. When the branches cross, a degenerate pair of $J = 0$ solutions appears, also labeled by a repective integer $n$. As $n$ increases, the mass of the rotating $J = 0$ solutions tends to the mass of the extremal RN black hole.

This intriguing branch structure for $Q < 0$ is also seen in Fig. 7, where the domain of existence of the EMCS solutions for $\lambda = 5$ is shown. While forming the boundary of the domain of existence, extremal EMCS solutions reside also deep within this domain along with the extremal static RN solution (black cross). Only the extremal $n = 1$ rotating $J = 0$ solutions

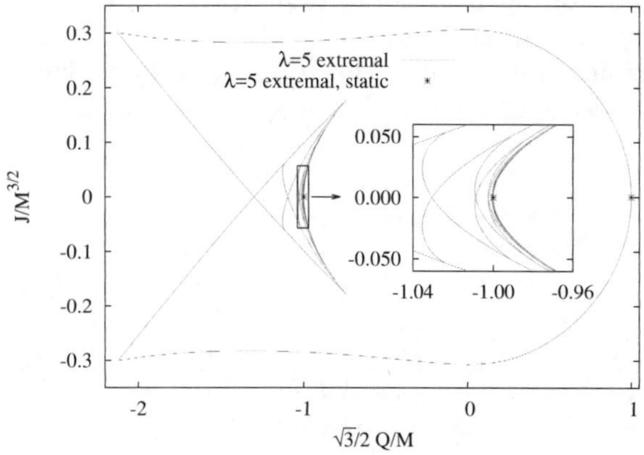


**Figure 7.** Domain of existence of $D = 5$ EMCS black holes: angular momentum vs charge (scaled with appropriate powers of the mass) for CS coupling $\lambda = 5$.

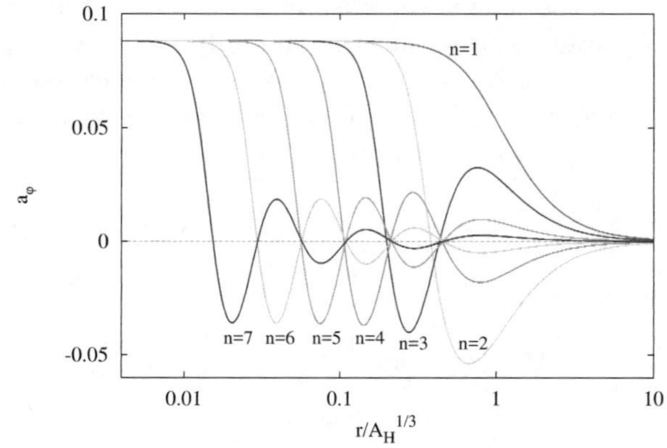


**Figure 8.** $D = 5$ EMCS black holes: Node structure of the gauge field function $a_\varphi$ ($\lambda = 5$, $n = 1, \ldots, 7$). The numbering includes the node at infinity.

are part of the boundary. The solutions with vanishing area represent the boundary solutions at the cusps at maximal $Q/M$.

When analyzing the branch structure one realizes, that the integer number $n$ labeling the branches corresponds to the number of nodes of the gauge field function $a_\varphi$ as illustrated in Fig. 8 for $n = 1, \ldots, 7$. While solutions with more than thirty nodes have been constructed in a systematic way, one is lead to conjecture, that there is in fact an infinity sequence of solutions, $n = 1, \ldots, \infty$. Note, that the drag function of the metric increases its nodes in an analogous manner. This sequence of radially excited rotating black holes is reminiscent of other physical systems with radial excitations, such as the hydrogen atom.

## 3. EMCS Solutions with AdS Asymptotics

Let us now include a negative cosmological constant, $\Lambda = -6/L^2$, with AdS length scale $L$. Then the solutions are no longer asymptotically flat but asymptotically AdS. The dS/AdS generalizations of the MP black holes are known in closed form [12, 13]. But the corresponding EMCS black holes could be obtained in closed form so far only for the case of gauged supergravity ($\lambda = 1$), where in a particular limit they are known to preserve some amount of supersymmetry [14, 6, 7].

### *3.1. Charged solutions*

We first address the question of how the presence of the negative cosmological constant affects the properties of the rotating black holes discussed above. The interesting new features present for the asymptotically flat solutions are basically retained in the presence of the cosmological term. In particular, when the CS coupling is sufficiently large, there appear counterrotating black holes. The black holes are no longer uniquely specified by their global charges. Instead an analogous branch structure arises, where the solution branches can be labeled by the node number of the magnetic gauge potential function.

For CS coupling in the vicinity of the supergravity value, we demonstrate in Fig. 9 as an example the $\lambda$-dependence of the area vs the charge and temperature. The extremal solutions form the lower boundary ($T_H = 0$). The charge symmetry breaking by the CS term is again clearly visible, with the zero area solutions only present for negative $Q$.

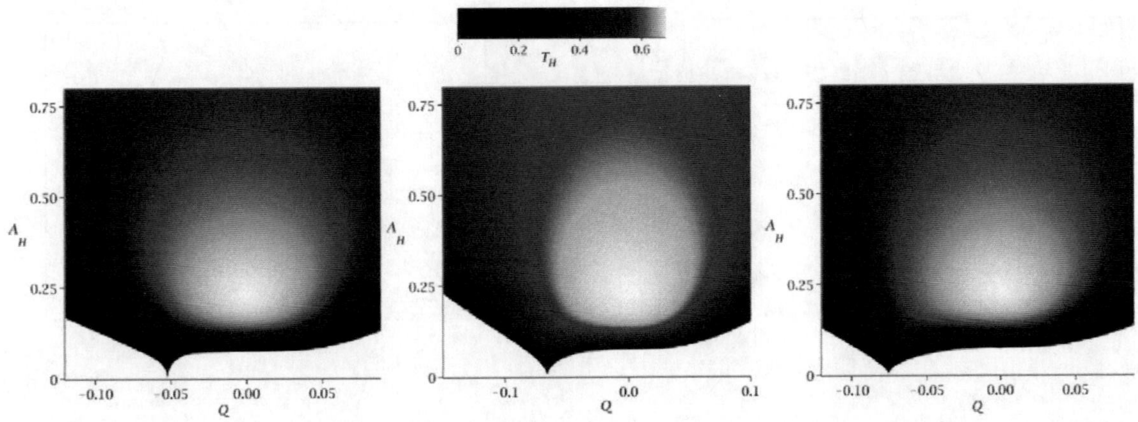


**Figure 9.** $D = 5$ EMCS AdS black holes: horizon area vs charge at different temperatures ($J = 0.00296$, $L = 1$) for $\lambda = 0.5$ (a), $\lambda = 1$ (b), and $\lambda = 1.5$ (c).

A new feature arises for small CS coupling $\lambda$. Here a near-horizon analysis shows, that the two types of branches (RN and MP for $\Lambda = 0$) no longer cross. This leads to a gap in the set of regular extremal black hole solutions, where the set of non-extremal solutions is limited by the so-called gap set, which seems to consist of singular solutions (as based on an extrapolation of the properties of the non-extremal solutions).

### *3.2. Magnetized solutions*

Let us now include a new physical feature, namely a new parameter $c_m$ of the AdS solutions associated with the (finite) magnitude of the magnetic potential at infinity, i.e., for $r \to \infty$: $a_\varphi \to c_m$. This parameter therefore determines the magnetic flux through the base space $S^2$ of the $S^1$ fibration of the $S^3$ (see Eqs. 4-5)

$$\Phi_m = \frac{1}{4\pi} \int_{S^2_\infty} \mathcal{F} = -\frac{1}{2} c_m. \tag{8}$$

For simplicity we first address the case of static EM solutions, since, acting like a box, the AdS background also allows for static solitonic EM solutions [15]. In particular, by considering magnetic fields only, one obtains regular EM solitons in $D = 5$ (and higher odd dimension), which are characterized by the magnetic flux parameter $c_m$ [16]. Interestingly, there is a maximal value of $c_m$, where such solitons exist.

By imposing a regular horizon, static EM black hole solutions arise, which can be classified into two types. Figs. 10 and 11 show their horizon area and mass versus their temperature. In type I black holes the horizon area and the mass increase with the temperature. In contrast, in type II black holes both decrease as the temperature increases. Type II black holes can be deformed continuously into solitons, when the horizon size approaches zero. Thus they exist only below the maximal value of $c_m$ for solitons. Type I black holes on the other hand exist also for large values of $c_m$, but they become singular in the limit of vanishing horizon size.

Clearly the above black holes represent new families of static EM AdS black holes, whose properties differ from those of the known RN AdS black holes. This suggests that similar magnetized EMCS AdS solutions could also exist. However, in the EMCS case solutions with a magnetic field must necessarily rotate. Indeed rotating EMCS generalizations of the above soliton and black hole solutions exist [17]. These solutions then also carry an electric charge,

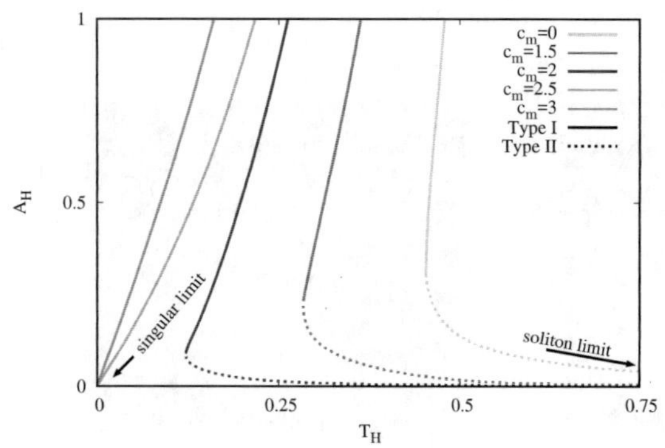


**Figure 10.** Static magnetized AdS EM black holes: area vs temperature (magnetic flux parameter $c_m = 0$ to 3).

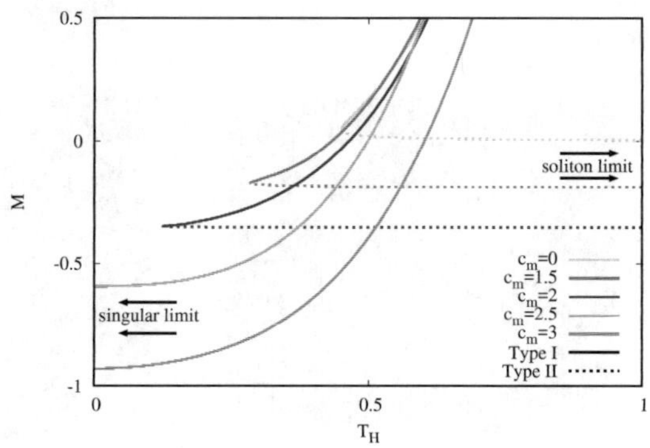


**Figure 11.** Same as Fig. 10 for the mass vs the temperature. Singular and soliton limits are indicated.

where we should now distinguish between the ordinary charge, the Page charge, and the $R$ charge

$$Q^{(R)} = -\frac{1}{2}\int_{S^3_\infty}\left(\tilde{\mathcal{F}} + \frac{2\lambda}{3\sqrt{3}}\mathcal{A}\wedge\mathcal{F}\right), \tag{9}$$

since the second term now contributes, and its prefactor differs for these three types of charges.

### *3.3. Magnetized squashed solutions*

Let us now consider a last twist w.r.t. the plethora of EMCS solutions by considering solutions which are asymptotically only locally AdS. Thus we impose that the magnetized rotating EMCS solutions should asymptotically approach not a round $S^3$ sphere but instead a squashed sphere. As discussed in [18] the boundary metric can then be expressed as

$$ds^2_B = L^2 d\Omega^2_{(v)} - dt^2, \qquad d\Omega^2_{(v)} = \frac{1}{4}\left(d\theta^2 + \sin^2\theta d\phi^2 + (d\psi + v\cos\theta d\phi)^2\right), \tag{10}$$

where $v$ is a control parameter. Clearly, for $v = 1$ the $S^3$ sphere becomes round, whereas for $v = 0$ the solutions tend to AdS black strings and vortices, their boundary corresponding to $S^2 \times S^1$.

Then for a given $\lambda$ a family of squashed magnetized soliton solutions arises smoothly from the AdS vacuum, which are labeled by the parameter $v$ and satisfy the relation $J = \Phi_m Q^{(R)}$. In the case of gauged supergravity ($\lambda = 1$) these solitons retain some amount of supersymmetry [18]. In fact, the properties of these squashed susy solitons can be expressed simply in terms of the squashing parameter $v$.

As one might expect, the solitons are related to a new class of squashed magnetized black holes. In the general case, these black holes are characterized by their mass, charge, angular momentum and magnetic flux parameter. In the limit of vanishing horizon radius solitons can be approached for an appropriate parameter choice. Families of black hole solutions for the case of gauged supergravity are exhibited in Figs. 12 and 13.

Generically the black holes possess an extremal limit with a finite horizon area. Among these extremal black holes a particular family stands out: the supersymmetric black holes. For these solutions the trace of the boundary stress tensors vanishes, yielding the condition $c_m = \pm\frac{L}{\sqrt{3}}(1 - v^2)$. Like the susy solitions these susy black holes can be characterized by the squashing parameter $v$.

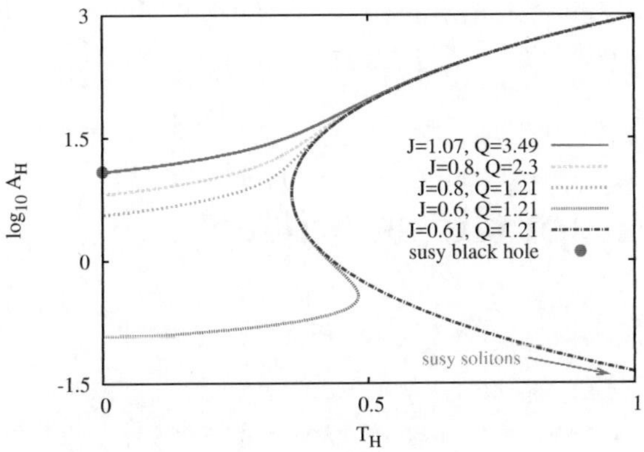


**Figure 12.** Squashed magnetized EMCS AdS black hole solutions: area vs temperature ($\lambda = 1$, $L = 1$, $v = 1.65$, $c_m = -1$).

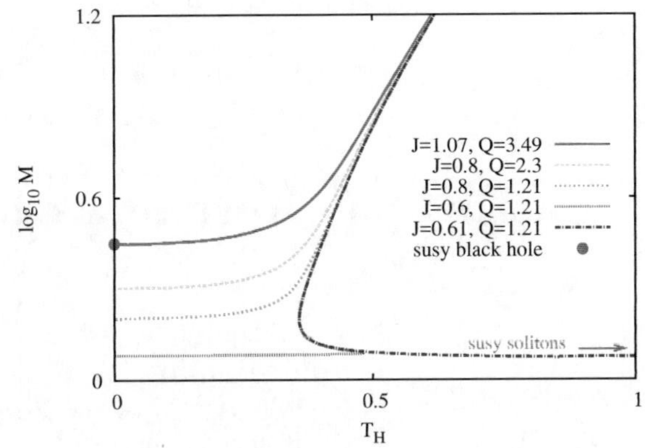


**Figure 13.** Same as Fig. 12 for the mass vs the temperature. The red dot marks a susy black hole.

This observation then leads to the following new picture for the classes of supersymmetric EMCS AdS black holes. Besides the previously known family of Gutowski-Reall black holes there exists a new family of squashed magnetized supersymmetric black holes. In fact, the new susy squashed magnetized black holes intersect the Gutowski-Reall black holes at a critical configuration, where the squashing and magnetization vanish, i.e., $v = 1$ and $c_m = 0$.

Obviously, there are still many unanswered questions, and many avenues are open awaiting further investigations. This holds, in particular, w.r.t. the relevance of the above configurations in an AdS/CFT context.

*Acknowledgments*
We gratefully acknowledge support by the DFG Research Training Group 1620 "Models of Gravity". E. R. acknowledges funding from the FCT-IF programme. F. N.-L. acknowledges funding from Complutense University - Santander under project PR26/16-20312.

## References

[1] Myers R C and Perry M J 1986 *Annals Phys.* **172** 304
[2] Kunduri H K and Lucietti J 2013 *Living Rev. Rel.* **16** 8
[3] Blázquez-Salcedo J L, Kunz J and Navarro-Lérida F 2013 *Phys. Lett.* B **727** 340
[4] Blázquez-Salcedo J L, Kunz J and Navarro-Lérida F 2014 *Phys. Rev.* D **89** 024038
[5] Breckenridge J C, Myers R C, Peet A W and Vafa C 1997 *Phys. Lett.* B **391** 93
[6] Cvetic M, Lu H and Pope C N 2004 *Phys. Lett.* B **598** 273
[7] Chong Z W, Cvetic M, Lu H and Pope C N 2005 *Phys. Rev. Lett.* **95** 161301
[8] Gauntlett J P, Myers R C and Townsend P K 1999 *Class. Quant. Grav.* **16** 1
[9] Kunz J and Navarro-Lérida F 2006 *Phys. Rev. Lett.* **96** 081101
[10] Blázquez-Salcedo J L, Kunz J, Navarro-Lérida and Radu E 2014 *Phys. Rev. Lett.* **112** 011101
[11] Blázquez-Salcedo J L, Kunz J, Navarro-Lérida and Radu E 2015 *Phys. Rev.* D **92** 044025
[12] Hawking S W, Hunter C J and Taylor-Robinson M M 1999 *Phys. Rev.* D **59** 064005
[13] Gibbons G W, Lu H, Page D N and Pope C N 2005 *J. Geom. Phys.* **53** 49
[14] J. B. Gutowski and H. S. Reall, Gutowski J B and Reall H S 2004 *JHEP* **0402** 006
[15] Hedeiro C and Radu E 2015 *Phys. Lett.* B **749**, 393
[16] Blázquez-Salcedo J L, Kunz J, Navarro-Lérida and Radu E 2016 *Entropy* **18** 438
[17] Blázquez-Salcedo J L, Kunz J, Navarro-Lérida and Radu E 2017 *Phys. Lett.* B **771** 52
[18] Cassani D and Martelli D 2014 *JHEP* **1408** 044

IOP Conf. Series: Journal of Physics: Conf. Series **942** (2017) 012004 doi:10.1088/1742-6596/942/1/012004

# Testing the Kerr black hole hypothesis with RELXILL_NK

**Cosimo Bambi**

Center for Field Theory and Particle Physics and Department of Physics, Fudan University, 200438 Shanghai, China

Theoretical Astrophysics, Eberhard-Karls Universität Tübingen, 72076 Tübingen, Germany

E-mail: bambi@fudan.edu.cn

**Abstract.** Astrophysical black hole candidates are thought to be the Kerr black holes of general relativity. However, macroscopic deviations from the Kerr background are predicted by a number of scenarios beyond Einstein's gravity. X-ray reflection spectroscopy can be a powerful tool to probe the strong gravity region of these objects and test the Kerr black hole hypothesis. Here I briefly review the state of the art of this line of research and I present some constraints on possible deviations from the Kerr metric obtained with the new X-ray reflection model RELXILL_NK and XMM-Newton, NuSTAR, and Swift data of the supermassive black hole in 1H0707-495.

## 1. Introduction

The theory of general relativity was proposed by Einstein in 1915. In the past sixty years, the theory has been extensively tested in weak gravitational fields, mainly with experiments in the Solar System and observations of binary pulsars, and current data well agree with the theoretical predictions [1]. However, there are a number of scenarios beyond Einstein's gravity that have the same predictions for weak fields and differ when gravity becomes strong. The ideal laboratory to test strong gravity is the spacetime around astrophysical black holes [2].

In 4-dimensional Einstein's gravity, the only stationary, asymptotically flat, regular on and outside the event horizon (i.e. without singularities and closed time-like curves), vacuum solution is the Kerr metric [3], which is completely characterized by two parameters, namely the black hole mass $M$ and the black hole spin angular momentum $J$. In Boyer-Lindquist coordinates, the line element reads ($G_{\rm N} = c = 1$)

$$\begin{aligned} ds^2 &= -\left(1 - \frac{2Mr}{\Sigma}\right) dt^2 + \frac{\Sigma}{\Delta} dr^2 + \Sigma\, d\theta^2 \\ &\quad + \left(r^2 + a^2 + \frac{2a^2 Mr \sin^2\theta}{\Sigma}\right) d\phi^2 - \frac{4aMr\sin^2\theta}{\Sigma} dt\, d\phi \,, \end{aligned} \qquad (1)$$

where $a = J/M$, $\Sigma = r^2 + a^2\cos^2\theta$, and $\Delta = r^2 - 2Mr + a^2$. $M$ and $J$ cannot be arbitrary and must satisfy the condition for the existence of the even horizon $|a_*| \le 1$, where $a_* = J/M^2$ is the dimensionless spin parameter.

The spacetime metric around astrophysical black holes formed from the gravitational collapse of massive bodies is expected to be well approximated by the Kerr solution [2]. Initial deviations

IOP Conf. Series: Journal of Physics: Conf. Series **942** (2017) 012004 doi:10.1088/1742-6596/942/1/012004

from the Kerr metric are quickly radiated away with emission of gravitational waves after the formation of the horizon. The equilibrium electric charge is extremely small and completely negligible for the spacetime geometry. The presence of accretion disks can be usually ignored. In the end, macroscopic deviations from the Kerr metric are only possible in the presence of new physics.

## 2. How can we test the Kerr black hole hypothesis?

There are two main approaches to test the Kerr nature of astrophysical black holes: with electromagnetic radiation [4, 5] and with gravitational waves [6]. Each method has its own advantages and disadvantages, and the two techniques are actually complementary, because deviations from standard predictions may lead to observational effects in one of the two approaches and not in the other and vice versa.

The electromagnetic approach is sensitive to the motion of the gas particles in the accretion disk and to the propagation of the photons from the point of emission in the strong gravity region to the point of detection in the flat faraway region. There are two natural strategies to test the Kerr black hole hypothesis with electromagnetic radiation [2]. In the so-called top-down approach, we consider some alternative theory of gravity in which black holes are not described by the Kerr metric and we check whether astrophysical data prefer the Kerr metric of Einstein's gravity or the non-Kerr metric of that specific alternative theory of gravity. This strategy has two main problems. First, there are a large number of alternative theories of gravity, and none seems to be more motivated than others, so we should repeat the analysis for every theory. Second, rotating black hole solutions in alternative theories of gravity are known only in quite exceptional cases, while the non-rotating or slow-rotating solutions are not very useful to test astrophysical black holes because the spin plays an important role in the properties of the electromagnetic radiation.

In the bottom-up approach, we employ a phenomenological test-metric in which possible deviations from the Kerr solution are quantified by one or more "deformation parameters". We measure the values of these deformation parameters with astrophysical data and we check whether they vanish, as it is required to recover the Kerr metric of Einstein's gravity.

## 3. X-ray reflection spectroscopy

X-ray reflection spectroscopy refers to the study of the reflection spectrum of accretion disks around black holes (see, e.g., [2] and references therein). Within the disk-corona model, a black hole is surrounded by a geometrically thin and optically thick accretion disk. The disk radiates like a blackbody locally and as a multi-color blackbody when integrated radially. The "corona" is a hotter, usually optically thin, electron cloud enshrouding the disk. Its exact geometry is currently unknown: it may be the base of the jet, the atmosphere above the disk, some kind of accretion flow between the disk and the central black holes, etc. Because of inverse Compton scattering of thermal photons from the disk off free electrons in the corona, the latter becomes an X-ray source with a power-law spectrum. Some X-ray photons from the corona hit the accretion disk, producing a reflection component with some fluorescent emission line.

The most prominent feature of the reflection spectrum is usually the iron K$\alpha$ line, which is at 6.4 keV in the case of neutral and weakly ionized iron and shifts up to 6.97 keV for H-like iron ions. While this line is very narrow in the rest-frame of the emitter, the one observed in the spectrum of black holes is broad and skewed due to special and general relativistic effects (gravitational redshift, Doppler boosting, light bending) occurring in the strong gravity region. In the presence of the correct astrophysical model, the analysis of the iron K$\alpha$ line and of the whole reflection spectrum can be a powerful tool to probe the spacetime around the black hole. The technique was originally developed assuming the Kerr metric to measure the black hole spin

IOP Conf. Series: Journal of Physics: Conf. Series **942** (2017) 012004 doi:10.1088/1742-6596/942/1/012004

parameter $a_*$ [7, 8, 9]. More recently, it was extended to generic metrics to test the Kerr black hole hypothesis [10, 11, 12, 13, 14, 15, 16].

## 4. RELXILL_NK

RELXILL is currently the most advanced model to describe the X-ray reflection spectrum of accretion disks in the Kerr spacetime [17, 18, 19]. Its extension to non-Kerr spacetimes is RELXILL_NK and was presented in [20]. RELXILL_NK currently employ the Johannsen metric with the deformation parameter $\alpha_{13}$ [21], but can easily be extended to any stationary, axisymmetric, and asymptotically flat black hole spacetime. In Boyer-Lindquist coordinates, the line element reads

$$\begin{aligned} ds^2 &= -\frac{\Sigma\left(\Sigma - 2Mr\right)}{A^2}\,dt^2 + \frac{\Sigma}{\Delta}\,dr^2 + \Sigma\,d\theta^2 + \frac{\left[\left(r^2+a^2\right)^2\left(1+\delta\right)^2 - a^2\Delta\sin^2\theta\right]\Sigma\sin^2\theta}{A^2}\,d\phi^2 \\ &\quad - \frac{2a\left[2Mr + \delta\left(r^2+a^2\right)\right]\Sigma\sin^2\theta}{A^2}\,dt\,d\phi\,, \end{aligned} \tag{2}$$

where $A = \Sigma + \delta\left(r^2 + a^2\right)$ and $\delta = \alpha_{13}\left(M/r\right)^3$. For $\alpha_{13} = 0$, we exactly recover the Kerr metric in (1). $\alpha_{13}$ cannot be arbitrary; it must satisfy the condition

$$\alpha_{13} \geq -\left(1 + \sqrt{1 - a_*^2}\right)^3\,, \tag{3}$$

in order to have a regular exterior region (no singularities or closed time-like curves).

## 5. Observational constraints from 1H0707-495

The study reported in Ref. [22] is the first attempt to test the Kerr nature of astrophysical black holes with real X-ray data. As source, we considered the supermassive black hole in the galaxy 1H0707-495: its spectrum is characterized by significant edge features, which are commonly interpreted as an extremely strong reflection component. This suggests that such a source can be a good candidate for testing the Kerr black hole hypothesis with RELXILL_NK.

There are several observations of 1H0707-495 with XMM-Newton, NuSTAR, and Swift. For XMM-Newton, we considered the 98 ks observation of 2011, which corresponds to the lowest flux state of 1H0707-495 and has been investigated by several authors. We fitted the data with two models

$$\begin{aligned} &\text{Model 1}: \quad \text{TBabs} * (\text{relxill\_nk} + \text{diskbb})\,, \\ &\text{Model 2}: \quad \text{TBabs} * (\text{relxill\_nk} + \text{relxill\_nk})\,. \end{aligned} \tag{4}$$

TBABS takes the galactic dust absorption into account. RELXILL_NK is the disk's reflection spectrum. DISKBB is the thermal component of a Newtonian disk. See [22] and reference therein for more details. Fig. 1 shows the constraints on the spin parameter $a_*$ and the deformation parameter $\alpha_{13}$ from Model 1 (left panel) and Model 2 (right panel). The red, green, and blue lines indicate, respectively, the 68%, 90%, and 99% confidence level curves.

Concerning NuSTAR and Swift data, there are three separated observations of 1H0707-495 with NuSTAR in 2014 and each observation has a simultaneous snapshot of Swift. In our study we excluded the second Swift observation because it was taken during an anomaly period of this mission. We employed the following model (for more details, see Ref. [22])

$$\text{Model 3}: \quad \text{TBabs} * \text{relxill\_nk}\,. \tag{5}$$

Fig. 2 shows the constraints on $a_*$ and $\alpha_{13}$ obtained from the NuSTAR+Swift data. The green and blue lines indicate, respectively, the 90% and 99% confidence level curves, while we do not report the 68% confidence level curve because it is too thin to be plotted.

IOP Conf. Series: Journal of Physics: Conf. Series **942** (2017) 012004 doi:10.1088/1742-6596/942/1/012004

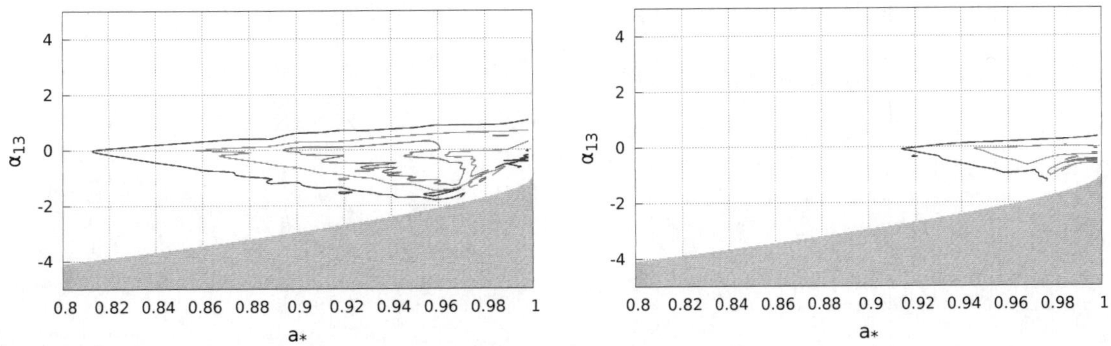


**Figure 1.** Constraints on the spin parameter $a_*$ and the Johannsen deformation parameter $\alpha_{13}$ from the XMM-Newton data of 2011: Model 1 (left panel) and Model 2 (right panel). The red, green, and blue lines indicate, respectively, the 68%, 90%, and 99% confidence level curves for two relevant parameters. The grayed region is outside the range prescribed for $\alpha_{13}$ in Eq. (3) and therefore is ignored in our study. From Ref. [22].

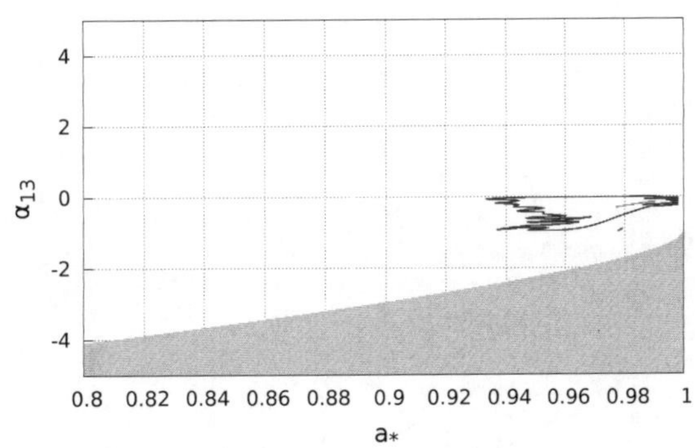


**Figure 2.** As in Fig. 1 for Model 3 and the data NuSTAR+Swift. The 68% confidence level curve is too thin to be plotted. From Ref. [22].

## 6. Opportunities with future X-ray missions

In Ref. [20], simulations were performed to test the capabilities of RELXILL_NK in analyzing observations from NuSTAR (as an example of a present instrument) and LAD/eXTP [23] (as an example of the next generation of X-ray missions). The resulting constraints are shown in Figs. 3, 4, and 5. In every figure, the left panel is for NuSTAR, the right panel for LAD/eXTP. In Fig. 3, the reference model is a Kerr black hole with the spin parameter $a_* = 0.8$ and the inclination angle $i = 30°$. In Fig. 4, we considered a non-Kerr black hole with the deformation parameter $\alpha_{13} = -2$, the spin parameter $a_* = 0.8$, and the inclination angle $i = 30°$. Fig. 5 shows the constraints obtained from a simulated observation of a Kerr black hole with the spin parameter $a_* = 0.8$ and the inclination angle $i = 80°$.

The constrains obtained simulating data with NuSTAR in Ref. [20] are comparable with those obtained in the analysis of 1H0707-495 in Ref. [22]. We can thus expect that the constraining power of the next generation of X-ray mission is significantly better, as we find in our simulations.

IOP Conf. Series: Journal of Physics: Conf. Series **942** (2017) 012004 doi:10.1088/1742-6596/942/1/012004

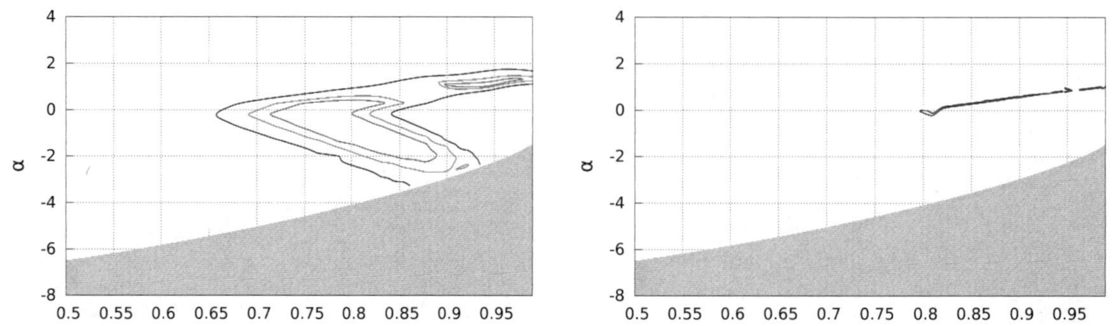


**Figure 3.** 68%, 90%, and 99% confidence level curves for the spin parameter $a_*$ and the deformation parameter $\alpha_{13}$ from a simulated observation of a bright stellar-mass black hole with NuSTAR (left panel) and LAD/eXTP (right panel). The spacetime metric of the simulation has $\alpha_{13} = 0$ (Kerr) and $a_* = 0.8$; the viewing angle is $i = 30°$. From Ref. [20].

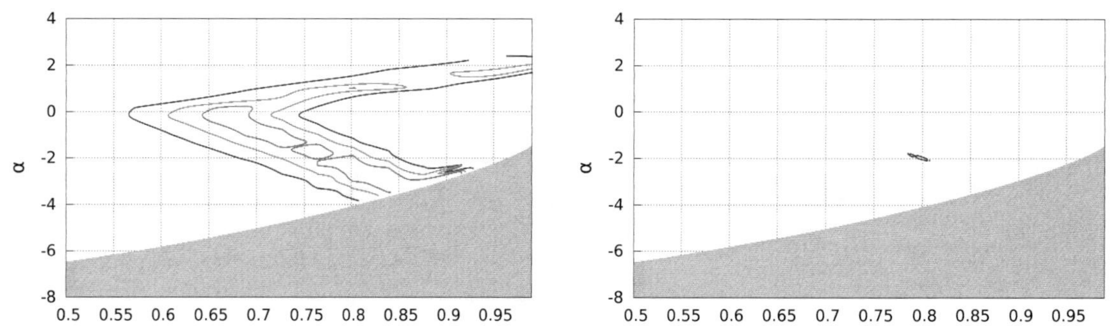


**Figure 4.** As in Fig. 3 for $\alpha_{13} = -2$, $a_* = 0.8$, and $i = 30°$. From Ref. [20].

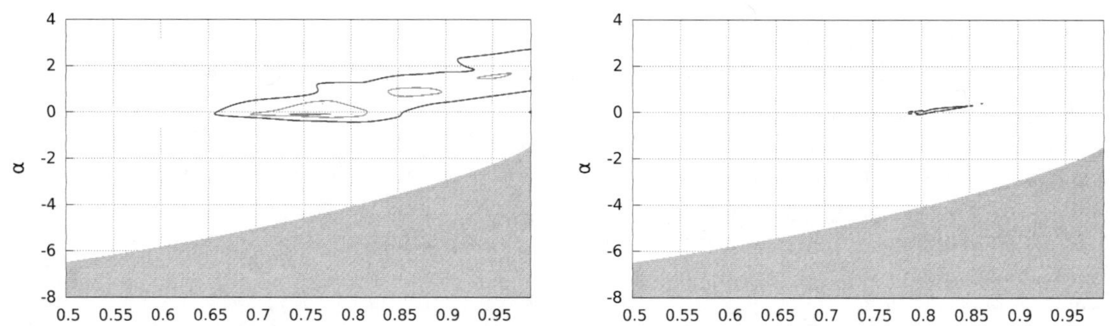


**Figure 5.** As in Fig. 3 for $\alpha_{13} = 0$ (Kerr), $a_* = 0.8$, and $i = 80°$. From Ref. [20].

## 7. Concluding remarks

RELXILL_NK is the extension of the X-ray reflection model RELXILL to probe the metric around astrophysical black holes and test the Kerr black hole hypothesis using X-ray reflection spectroscopy [20]. In Ref. [22], we employed for the first time the new model to analyze XMM-Newton, NuSTAR, and Swift data of the supermassive black hole in 1H0707-495. Our results are shown in Figs. 1 and 2, and are consistent with the assumption that the metric around the supermassive black hole in 1H0707-495 is described by the Kerr solution, as expected in Einstein's gravity. We are currently working to test the Kerr metric with other sources (both stellar-mass and supermassive black hole candidates) as well as different kinds of deformations from the Kerr solution. Simulations show that significantly better constraints can be obtained with the next generation of X-ray missions [20].

## Acknowledgments

This work was supported by the National Natural Science Foundation of China (Grant No. U1531117), Fudan University (Grant No. IDH1512060), and the Alexander von Humboldt Foundation.

## References

[1] C. M. Will, Living Rev. Rel. **17**, 4 (2014) [arXiv:1403.7377 [gr-qc]].
[2] C. Bambi, *Black Holes: A Laboratory for Testing Strong Gravity* (Springer Singapore, 2017), doi:10.1007/978-981-10-4524-0
[3] P. T. Chrusciel, J. L. Costa and M. Heusler, Living Rev. Rel. **15**, 7 (2012) [arXiv:1205.6112 [gr-qc]].
[4] C. Bambi, Rev. Mod. Phys. **89**, 025001 (2017) [arXiv:1509.03884 [gr-qc]].
[5] C. Bambi, J. Jiang and J. F. Steiner, Class. Quant. Grav. **33**, 064001 (2016) [arXiv:1511.07587 [gr-qc]].
[6] N. Yunes and X. Siemens, Living Rev. Rel. **16**, 9 (2013) [arXiv:1304.3473 [gr-qc]].
[7] A. C. Fabian, K. Iwasawa, C. S. Reynolds and A. J. Young, Publ. Astron. Soc. Pac. **112**, 1145 (2000) [astro-ph/0004366].
[8] L. W. Brenneman and C. S. Reynolds, Astrophys. J. **652**, 1028 (2006) [astro-ph/0608502].
[9] C. S. Reynolds, Space Sci. Rev. **183**, 277 (2014) [arXiv:1302.3260 [astro-ph.HE]].
[10] J. Schee and Z. Stuchlik, Gen. Rel. Grav. **41**, 1795 (2009) [arXiv:0812.3017 [astro-ph]].
[11] T. Johannsen and D. Psaltis, Astrophys. J. **773**, 57 (2013) [arXiv:1202.6069 [astro-ph.HE]].
[12] C. Bambi, Phys. Rev. D **87**, 023007 (2013) [arXiv:1211.2513 [gr-qc]].
[13] J. Jiang, C. Bambi and J. F. Steiner, Astrophys. J. **811**, 130 (2015) [arXiv:1504.01970 [gr-qc]].
[14] M. Zhou, A. Cardenas-Avendano, C. Bambi, B. Kleihaus and J. Kunz, Phys. Rev. D **94**, 024036 (2016) [arXiv:1603.07448 [gr-qc]].
[15] Y. Ni, M. Zhou, A. Cardenas-Avendano, C. Bambi, C. A. R. Herdeiro and E. Radu, JCAP **1607**, 049 (2016) [arXiv:1606.04654 [gr-qc]].
[16] C. Bambi, Z. Cao and L. Modesto, Phys. Rev. D **95**, 064006 (2017) [arXiv:1701.00226 [gr-qc]].
[17] T. Dauser, J. Garcia, J. Wilms, M. Bock, L. W. Brenneman, M. Falanga, K. Fukumura and C. S. Reynolds, Mon. Not. Roy. Astron. Soc. **430**, 1694 (2013) [arXiv:1301.4922 [astro-ph.HE]].
[18] J. Garcia, T. Dauser, C. S. Reynolds, T. R. Kallman, J. E. McClintock, J. Wilms and W. Eikmann, Astrophys. J. **768**, 146 (2013) [arXiv:1303.2112 [astro-ph.HE]].
[19] J. Garca *et al.*, Astrophys. J. **782**, 76 (2014) [arXiv:1312.3231 [astro-ph.HE]]
[20] C. Bambi, A. Cardenas-Avendano, T. Dauser, J. A. Garcia and S. Nampalliwar, Astrophys. J. **842**, 76 (2017) [arXiv:1607.00596 [gr-qc]].
[21] T. Johannsen, Phys. Rev. D **88**, 044002 (2013) [arXiv:1501.02809 [gr-qc]].
[22] Z. Cao, S. Nampalliwar, C. Bambi, T. Dauser and J. A. Garcia, arXiv:1709.00219 [gr-qc].
[23] S. N. Zhang *et al.* [eXTP Collaboration], Proc. SPIE Int. Soc. Opt. Eng. **9905**, 99051Q (2016) [arXiv:1607.08823 [astro-ph.IM]].

IOP Conf. Series: Journal of Physics: Conf. Series **942** (2017) 012005 doi:10.1088/1742-6596/942/1/012005

# OJ287 taken to pieces: the origin of a precessing and rotating jet

**S Britzen[1], C Fendt[2], G Witzel[3], S-J Qian[4], I N Pashchenko[5], O Kurtanidze[6,7], M Zajacek[1,8,10], G Martinez[3], V Karas[8], M Aller[9], H Aller[9], A Eckart[10,1], K Nilsson[11], P Arévalo[12], J Cuadra[13], and A Witzel[1]**

[1] Max-Planck-Institut für Radioastronomie, Auf dem Hügel 69, 53121 Bonn, Germany
[2] Max-Planck-Institut für Astronomie, Königstuhl, Heidelberg, Germany
[3] UCLA, Department of Physics and Astronomy, LA, CA 90095, USA
[4] National Astronomical Observatories, Chinese Academy of Sciences, Beijing 100012, China
[5] Astro Space Center, Lebedev Physical Institute, Russian Academy of Sciences
[6] Abastumani Observatory, Mt Kanobili, 0301 Abastumani, Georgia
[7] Engelhardt Astronomical Observatory, Kazan Federal University, Tatarstan, Russia
[8] Astronomical Institute, Academy of Sciences, Boční II 1401, 14131 Prague, Czech Republic
[9] University of Michigan, Ann Arbor, MI 48109, USA
[10] I. Physikalisches Institut der Universität zu Köln, Zülpicher Str. 77, 50937 Köln, Germany
[11] Tuorla Observatory, Department of Physics and Astronomy, University of Turku, 20500, Turku, Finland
[12] Instituto de Física y Astronomía, Facultad de Ciencias, Universidad de Valparaíso, Gran Bretaña No. 1111, Playa Ancha, 2360102 Valparaíso, Chile
[13] Instituto de Astrofísica, Pontificia Universidad Católica de Chile, 782-0436 Santiago, Chile

E-mail: `sbritzen@mpifr.de`

**Abstract.** OJ287 is the best candidate active galactic nucleus for hosting a supermassive binary black hole at very close separation, corresponding to the orbital period of the order of ~9 yr. We studied the pc-scale jet dynamics in 118 Very Long Baseline Array (VLBA) observations at 15 GHz covering the time between Apr. 1995 and Jan. 2017. To our knowledge, this is the first time, that the kinematics of the Blandford-Znajek jet (originating in the ergosphere of a rotating black hole) and jet sheath (originating from the accretion disk) are seen and traced in observations. We also find that the OJ287 radio jet is rotating and precessing. The jet dynamics as well as the flux-density light curves can be understood in terms of geometrical effects. A binary black hole model can explain the time scale of the precessing motion. Lense-Thirring precession of an accretion disc surrounding a single black hole is consistent with the time scale as well.

## 1. Introduction

Despite many high-resolution radio interferometric studies of the jets of active supermassive black holes in active galactic nuclei (AGN), still many questions remain to be answered. We do not know yet, whether the central black hole in an AGN is single or binary. We would like to know whether the jets are launched from the ergosphere of the black hole at the expense of its angular momentum (Blandford & Znajek 1977) or at the expense of orbital motion of the gas in the magnetised accretion disk (Blandford & Payne 1982), or a combination of both scenarios.

IOP Conf. Series: Journal of Physics: Conf. Series **942** (2017) 012005 doi:10.1088/1742-6596/942/1/012005

Many questions relate to the kinematics of the brightness enhancements which are typically seen in these jets on pc-scales and which separate with apparent superluminal speeds away from the central core. Some jets however, do solely, or in addition, show brightness enhancements which remain at similar core separations and only move with apparent subluminal speeds along the jet. The physical nature of those stationary components is still debated, e.g. as recollimation shocks (Daly & Marscher 1988).
OJ287 (z=0.306 (Stickel et al. 1989), a low-synchrotron peaked (LSP) BL Lac Object) is a prototypical AGN which allows us to answer some of these questions. In the following, we present an extract of our results which are described and discussed in full detail in Britzen et al. (2017).

### *1.1. OJ287 in a very brief summary*

OJ287 has been claimed to be the most promising AGN for hosting a supermassive binary black hole at the center. The basis for this claim is a light-curve observed in the optical V band which since 1890 shows repeated outbursts at ∼11.65 yr intervals (Sillanpää et al. 1988). Since the light curve during an outburst resembles the pattern of inflow of gas from an accretion disk to a supermassive black hole in a tidal perturbation, Sillanpää et al. (1988) proposed that OJ287 is a binary pair of supermassive black holes with an orbital period of 9 yr in the rest frame of OJ287. They also claimed that the light variations are related to tidally induced mass flows from accretion disks into black holes. Lehto & Valtonen (1996) explained well-defined sharp flares in the light-curve with a model in which a smaller black hole crosses the accretion disk of a larger black hole during the binary orbit of the black holes about each other. Further work by e.g., Lehto & Valtonen (1996), Valtonen & Wiik (2012b), Valtonen et al. (2016) have explored the binary BH nature of this AGN.
While the model by Valtonen et al. (2016) nicely explains the optical data, many still open questions remain. It has proved difficult to relate the flaring variability observed in the optical wavelength regime for OJ287 with the flaring variability observed in the radio on the one hand, and the morphological changes of the pc-radio radio jet on the other hand. In particular, an important question relates to the nature of the radio jet emission and the jet direction - does this really change erratically as the results by Agudo et al. (2012) suggest? Is the origin (or launching of the jet) affected by the plunging scenario (Valtonen et al. 2016)? Can the radio jet direction and its changes tell us more about the origin of the jet? We here summarize our results concerning the physical processes driving the OJ287-jet.
Throughout the paper we adopt the following parameters: a luminosity distance $D_L$=1.592 Gpc, at the source redshift of $z$=0.306, with cosmological parameters corresponding to a ΛCDM of the Universe with $\Omega_m$=0.3, Ω=0.7, and $H_0$=70 km $s^{-1}$ $Mpc^{-1}$. A proper motion of 1mas $yr^{-1}$ corresponds to an apparent superluminal speed of 14.7$c$. 1 mas = 4.521 pc.

## 2. The Observations

OJ287 was studied as part of the MOJAVE[1] (Monitoring Of jets in Active galactic nuclei with VLBA Experiments) survey (e.g., Hovatta et al. 2014, Lister et al. 2009, Lister et al. 2013). The presented study is based on a re-modeling and re-analysis of 118 VLBA data sets obtained at 15 GHz within the MOJAVE program (for details see Britzen et al. 2017). The data cover a time range between Apr 1995 and Jan 2017. In addition, we studied single-dish radio flux-density data at three frequencies (4.8, 8.0, 14.5 GHz) obtained within the University of Michigan (UMRAO) monitoring program. To analyze the optical data, we combined historical light-curve information with data taken by the Abastumani Observatory.

[1] www.physics.purdue.edu/astro/MOJAVE/

IOP Conf. Series: Journal of Physics: Conf. Series **942** (2017) 012005 doi:10.1088/1742-6596/942/1/012005

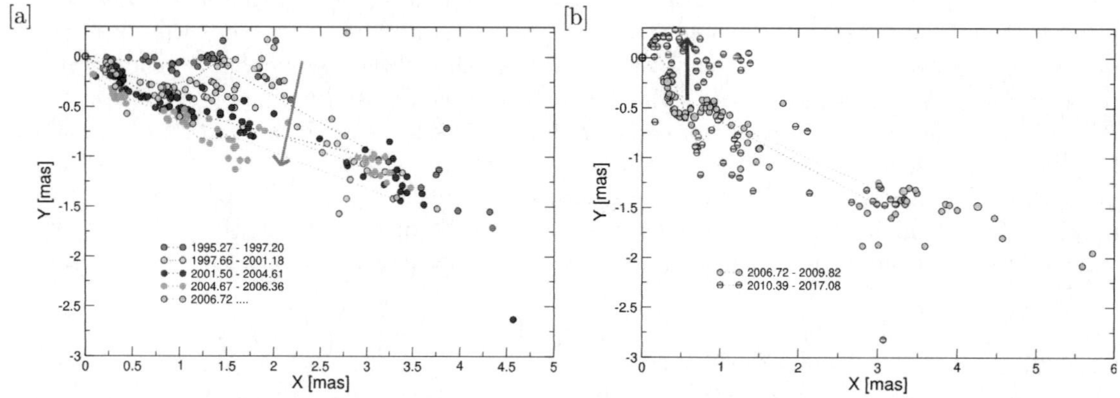


**Figure 1.** Jet precession. We plot all the (X, Y) data points of the 118 re-analyzed epochs. The origin (0, 0) is the VLBI core. The counter-jet components have been deliberately taken out. The sequence of the combined effect of rotation and precession starts in time in (a) with the earliest data (magenta). We indicate the motion and evolution in time with different colors. A significant shift in the plane of the sky is indicated with dotted lines. (b) is a continuation of (a). The last epoch in (a) displayed with green circles is the first phase in the plot on the right. From green the data continue with a "jump" to the red-striped circles in (b). In both plots, an arrow indicates the main direction of the projected jet precession motion. In (a) the dominant motion is downward. In (b) the dominant motion is upward.

## 3. The dynamics of the jet of OJ287

### *3.1. Precession*

We applied (circular) Gaussian components to the data to parametrize the jet emission along the jet. The "ridge line" connects all jet features modeled per epoch. We traced the evolution of these ridge lines in time. The ridge line evolution is shown Fig. 1 (a) and (b). To better indicate how the jet moves in the plane of the sky, we mark the dates by different colors. Fig. 1 (a) shows the earlier data (from 1995.27–2006.72) and (b) the ridge line evolution in later epochs (from 2006.72 –2017.08). The earlier data show how the jet in the plane of the sky moves downward (arrow in magenta indicates the direction of the motion). The later data show an upward trend (indicated by a blue arrow) which starts after epoch 2009.82. The data shown by the stripped circles (red-white) indicate the flip. We assume that the jet motion changed direction and allows a more direct view into the jet (smaller angle to the line of sight). The dominant effect visible in the data is that of a wandering jet in the sky. This is most likely the result of a *precessing* jet. The precessional motion of the jet can also be seen in Fig. 2 (b).

### *3.2. Apparent superluminal and subluminal jet components*

AGN jets reveal brightness enhancements along the jet. Tracing their motion through the epochs of observations allows us to study the dynamics of the jet. We find two types of motion of jet features in OJ287: components moving away from the core with apparent superluminal speeds (the standard *superluminal features* observed in quasar jets) and apparently subluminally moving components (in the direction along the jet) that remain at similar core separations - but - move perpendicular to the jet axis. These latter *"stationary"* components trace the rotation (and precession) of the jet. We show both types of components and their kinematic properties in Fig. 2 (a) and (b). Jet components denoted with **C** + number are those features moving with apparent superluminal speeds away from the core. Components denoted with **pattern** represent

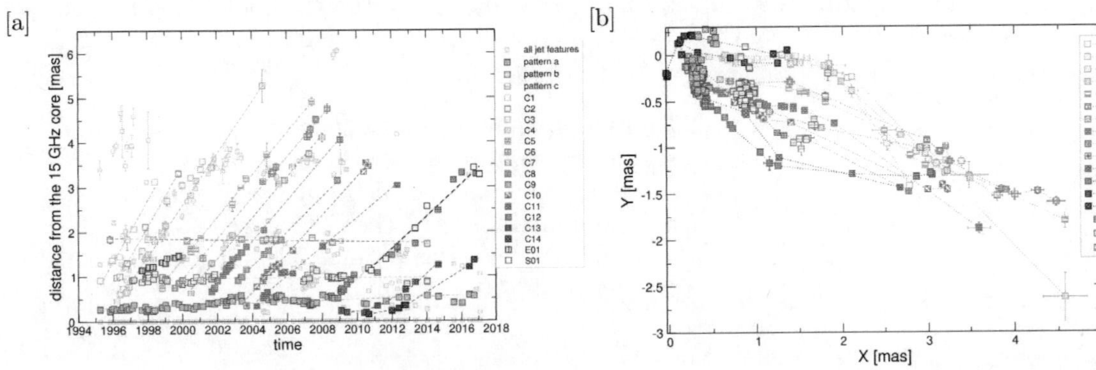


**Figure 2.** The distance between the 15 GHz core and the jet components plotted as function of time (a). Individual jet components are shown in different colors. For components up to **C12** the linear regression line is superimposed. For components **C13** and **C14** a quadratic fit has been applied (see text for details). (b) shows the components in xy-coordinates. Both figures have been prepared with the counter-jet components deliberately left out.

the stationary features moving with apparent subluminal speeds along the jet. From **C1** to **C11** the apparent speeds of the components are decreasing - almost continuously. Consequently, the viewing angle (assuming Doppler beaming) is increasing between 5.5 and 12.0 deg for the earliest appearing jet feature **C1** to **C11**. This can be interpreted as motion of the emitting source away from the line of sight. Apparent superluminal motion of 4-10$c$ is a common phenomenon in AGN. However, the physical nature of these components is still a matter of debate. We argue that the fast moving components in OJ287 represent the jet originating via the Blandford-Znajek-mechanism. The stationary features observed in the OJ287 jet trace the jet rotation as can be seen in Fig. 4. We propose that they represent the sheath of the jet originating from the accretion disk via the Blandford-Payne mechanism.

### *3.3. Jet rotation*

Evidence for a rotating jet is shown in Fig. 3. Consecutive jet component-paths (those with apparent superluminal motion) appear at systematically different places in the xy-plane. Tracing this movement of these four jet component-paths in Fig. 3 gives the signature of a rotation.

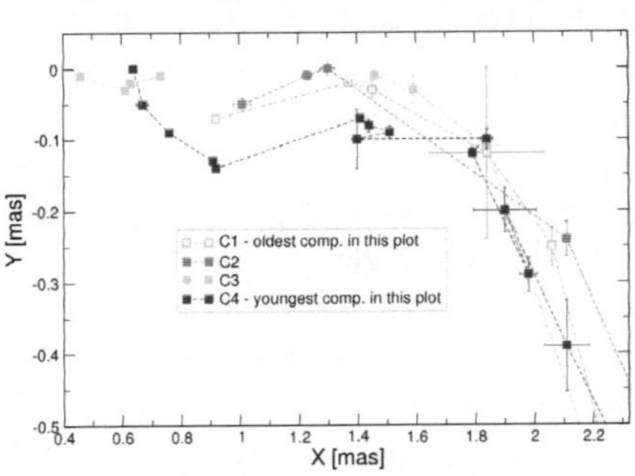


**Figure 3.** The paths of four consecutive moving jet components are shown, starting with **C1**. **C2** moves on a path displaced upward compared to **C1**, **C3** on a path upward of **C2**. The path of **C4** is clearly below those of the other three jet components. The displacement of the paths indicates rotation.

### *3.4. Rotation of the stationary jet features*

We find that those components that appear to move with apparent subluminal motion in the direction of the jet show significant motion perpendicular to the jet axis. In Fig. 4 we show,

IOP Conf. Series: Journal of Physics: Conf. Series **942** (2017) 012005 doi:10.1088/1742-6596/942/1/012005

that this motion is compatible with a rotation (perpendicular to the jet axis). This motion is consistent with the motion expected for the sheath of the jet originating in the accretion disk.

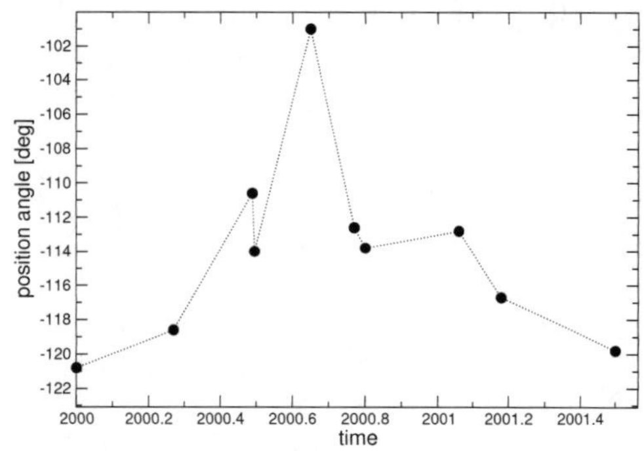


**Figure 4.** Jet component rotation. The stationary feature **a** reveals evidence for a rotation: the position angle increases up to a maximum value and then decreases again. This motion occurs on a time scale of roughly one year.

*3.5. OJ287 jet appears as a helix*

Collecting the data of one phase of the precession reveals a pattern resembling a *helix*. This is shown in Fig. 1 (b) with the circles filled in green. The signature of the helix results from an interplay of several directions of motions, in particular the components moving away from the core (with apparent superluminal speeds) and those components moving with apparent subluminal speeds in the direction along the jet but revealing significant motion perpendicular to the jet axis.

*3.6. Time-scales of VLBI kinematics match light-curve variability periodicities*

The short timescale of the rotation of the most prominent stationary component **a** coincides with the time-separation of the outbursts visually apparent in the radio light-curve (at 15 GHz) of about one year. The longer timescale (roughly 11-15 years) of half of the jet precession scale compares well with a period of enhanced activity in the single-dish radio flux-density monitoring data (UMRAO). This timescale compares quite nicely with the dominant periodicity quoted for the optical light-curve. It seems likely that the orbital motion of the rotating jet can - via Doppler beaming - explain the observed time-scales in the radio light-curve and in the optical light-curve. This is discussed in more detail in Britzen et al. (2017).

## 4. Conclusions

Our results for the OJ287 allow us to resolve some of the still outstanding questions related to the origin and dynamics of its jet and the physical nature of the radio variability of OJ287. We also offer an alternative explanation for the origin of the historical optical variability. In Britzen et al. (2017) we propose that the stationary components represent the sheath of the jet while the superluminal components represent the long sought-after Blandford-Znajek jet. To our knowledge, this is the first time both can be discriminated and investigated with regard to their kinematics. The physical origin of the radio variability of AGN has often been attributed to the accretion phenomenon. Matter is being accreted and part of this matter is ejected into the jet and appears as a jet component - this is a short summary of the generally accepted paradigm. We here show that the radio variability in OJ287 originates in the geometry of the system - the rotating jet determines the timescale of the rapid radio variability. The longer time scale variability seems to originate in the precession of the jet. We thus explain the radio variability in OJ287 in terms of deterministic, geometric processes.

Based on observations and application of the well-known black hole mass relations, a value of $4\times10^8 M_\odot$ has been derived (Liu & Wu 2002). Refering to this estimate and the mass

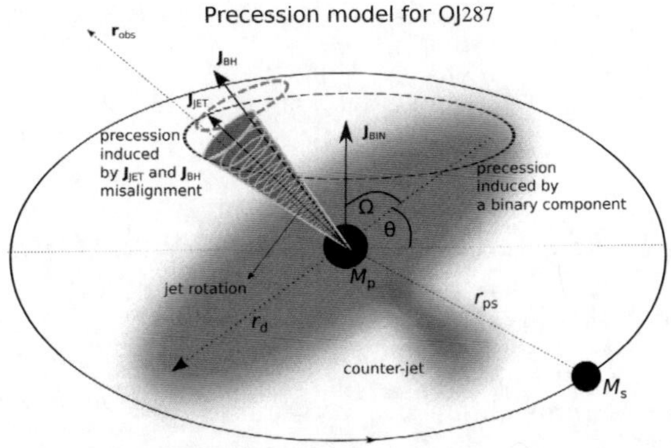


**Figure 5.** Sketch to illustrate two of the scenarios discussed in more detail in Britzen et al. 2017: 1. The Lense-Thirring precession due to the misalignment of angular momenta of the Kerr black hole and that of the accretion disc. 2. The disc-jet precession due to the torques induced by a secondary black hole.

determined within the Valtonen-model ($10^{10}M_{\odot}$), we discuss several possible scenarios in Britzen et al. (2017) to explain the observed jet precession (e.g., see Fig. 5). Both the Lense-Thirring precession and the precession due to a binary companion can explain the observed precession time scales. A binary black hole scenario - however - seems more likely. Our findings are consistent with a mass of $10^8M_{\odot}$, which can correspond to the mass of a single black hole or the total mass of the binary. A massive black hole of $10^{10}M_{\odot}$ is not required to explain the observed phenomena and time scales. OJ287 is a prototypical object. It is thus possible and likely that our findings have implications for the jets of BL Lac objects or AGN in general. This can significantly change our understanding of jet component motion, jet launching, radio variability, and the accretion process. Detailed modeling of the jet kinematics is in preparation.

### *4.1. Acknowledgments*
SB thanks the organizers of the KSM for a very inspiring meeting and great hospitality. This research has made use of data from the MOJAVE database that is maintained by the MOJAVE team (Lister et al., 2009, AJ, 137, 3718). The National Radio Astronomy Observatory is a facility of the National Science Foundation operated under cooperative agreement by Associated Universities, Inc.

## 5. References
[1] Agudo I, Marscher A P, Jorstad S G, et al. 2012 *ApJ* **747** 63
[2] Blandford R D and Znajek R L 1977 **MNRAS** *179* 433-456
[3] Blandford R D and Payne D G 1982 **MNRAS** *199* 883-903
[4] Britzen S, et al. 2017 subm. to **MNRAS**
[5] Daly R A and Marscher A P 1988 **ApJ** *334* 539-55
[6] Hovatta T, Pavlidou V, King O G, et al. 2014 *MNRAS* **439** 690
[7] Lehto H J and Valtonen M J 1996 **ApJ** *460* 207
[8] Lister M L, Homan D C, Kadler M, Kellermann K I, et al. 2009 *ApJ* **696** 22
[9] Lister M L, Aller M F, Aller H D, et al. 2013 *AJ* **146** 120
[10] Liu F K and Wu X B 2002 **A&A** *388* L48-L52
[11] Sillanpää A, Haarala S, Valtonen M J et al. 1988 *ApJ* **325** 628
[12] Stickel M, Fried J W, Kühr H 1989 *A&AS* **80** 103
[13] Valtonen M J and Wiik K 2012 *MNRAS* **421** 1861
[14] Valtonen M J, Zola S, Ciprini S, et al. 2016 *ApJL* **819** L37

IOP Conf. Series: Journal of Physics: Conf. Series **942** (2017) 012006 doi:10.1088/1742-6596/942/1/012006

# Illusion about weakness of gravity hides new way to unify gravity with particle physics

**Alexander Burinskii**

Theor. Phys. Lab., NSI, Russian Academy of Sciences, B. Tulskaya 52 Moscow 115191, Russia

E-mail: bur@ibrae.ac.ru

**Abstract.** Well known weakness of Gravity in particle physics is an illusion caused by underestimation of the role of spin in gravity. We argue that spin deforms space along with mass, and great spin/mass ratio of elementary particles shifts effective scale of gravitational interaction from Planck to Compton distances. It opens new way to unify gravity and quantum theory, which is achieved by a supersymmetric bag model, creating a flat Compton zone near the core of particle required for consistent work of quantum processes. Super-bag is naturally upgraded to Wess-Zumino supersymmetric QED model, forming a bridge to perturbative formalism of conventional QED.

## 1. Introduction

One of the requirement for unification of Quantum theory with Gravity is a field description of an extended particle, since the pointlike particles cannot give a stress-energy tensor which is necessary for the right side of the Einstein equations, $G_{\mu\nu} = 8\pi T_{\mu\nu}$. A perfect theory of unification was given by the old Kaluza-Klein model, where 4d Einstein-Maxwell equations ${}^4G_{\mu\nu} = 8\pi T_{\mu\nu}$, were created from 5-dimensional Einstein's tensor ${}^5G_{MN} = 0$, where the fifth component was generated by vector potential of ectromagnetic field $A^\mu$, forming a circle with cylindrical boundary conditions. The experimental invisibility of this extra dimension created idea to shift size of circle to Planck scale, where experiment cannot resolve this structure. The idea of extra dimension was inherited by string theory together with idea of invisibility at the Planck scale, and was generalized to many extra dimensions. It was revolutionary step since the pointlike and structureless quantum objects turned into extended stringlike objects, retaining close relationship with general relativity. Planck scale appeared "... as a rough estimate of the fundamental string length scale as well as the characteristic size of compact extra dimensions." [1] The additional arguments in favor of fundamentality of the Planck length was inclusion of the constant of gravitational interaction $G$ with remarkable 'naturalness' of the Planck units, $\bar{h} = c = G = 1$, and also the fact, that at the Planck length $l_P = (\hbar G/c^3)^{1/2})$, impact of gravitational Schwarzschild solution on space becomes commensurate with other interactions. As a result, during last century weakness of gravitational interaction and Planck scale of gravitational interaction were turned into an indisputable truth.

However, similarly to problems in Kaluza-Klein theory, the problems with unification appeared in superstring theory, and *"...Since 1974 superstring theory stopped to be considered as particle physics... "* and *"... a realistic model of elementary particles still appears to be a distant dream ... "*, J. Schwarz [2]. However science moves, and about half century ago Kerr

IOP Conf. Series: Journal of Physics: Conf. Series **942** (2017) 012006 doi:10.1088/1742-6596/942/1/012006

obtained a new gravitational solution for a spinning black hole (BH) [3]. It was obtained, the spinning and charged Kerr-Newman (KN) solution has strong influence on space not on the expected Planck scale, but on the much lover energy scale . In particular, it was obtained in [4, 5, 6] that the KN solution with parameters of an electron deforms space topologically at the Compton distance, which exceeds the Planck length on 22 orders.

The zone of space distortion depends on spin/mass ratio $a = J/m$, showing that spin deforms space along with mass, and it should be taken into account at estimation of the gravitational influence.

Indeed, formation of BH horizons is related with gravitational effect of "frame-dragging" [7], which is a local deformation of the light cone system. In the Schwarzschild solution, BH horizons are formed by a centrally-symmetric frame-dragging, while in the Kerr and KN solutions frame-dragging forms a vortex, which cannot be removed by a global Lorentz transformations, and was called as a Lense-Thirring effect. One can say that *spin is gravitating,* but its action on the metric is different. Contrary to spherically symmetric Schwarzschild solution with mass $m$, which distorts space in zone $r_g \sim 2m$, the Kerr spinning solution with angular momentum $J$ and mass $m$ distorts space in zone

$$r_K \sim J/m, \tag{1}$$

which becomes very essential for particle physics, because the spin/mass ratio of particles is typically about $\approx 10^{20} - 10^{22}$. [1]

So large extension of the zone of gravitational influence, about 20 orders for typical spinning particles shows that due to energy of spin influence of gravity on space-time is shifted from Planck scale to Compton distances $r_K$, just to zone of action of quantum theory. Because of the huge spin/mass ratio, gravitational influence cannot be considered as weak.

It's natural to compare this situation with Cosmology, where nobody says that gravity is weak because of giant masses. In the same way, Gravity is not weak in particle physics, because of the giant energy contribution from spinning particles.

The caused by spin effective deformation of space at Compton scale creates principally new situation, in which gravity and quantum theory become equally strong in vicinity of the Compton region, and should be considered on equal footings. Contrary to the old concept in which gravity is subordinated to quantum theory and must be quantized, rank of gravity is raised to theory of background space which may be theory of the deepest level. This requires a new conceptual approach to the problem of unification of quantum theory and gravity, and we present here a way to resolution of the conflict between these theories without quantization of gravity or modifications of the Einstein-Maxwell equations. We show that this remarkable mechanism is SUPERSYMMETRY which separates inner quantum region from gravity. We use a nonperturbative supersymmetric Higgs field model forming the super-bag, in which the region inside the bag is regularized by supersymmetric vacuum state with broken gauge symmetry. Inside the bag, gravitational and EM field acquire a mass and fall off, while the external space is retained as the exact KN solution.

The based on supersymmetric Higgs (or Landau-Ginzburg) field model nonperturbative solution has remarkable relationships with Wess-Zumino SuperQED model which gives a link to perturbative technics of the conventional QED. Here we discuss some features of this approach. For more details see [8, 9, 10, 11].

[1] We use the 'natural' (Planck) units $\hbar = c = G = 1$, in which Planck's mass, length and energy are $M_P = l_P = E_P = 1$. In this system the energy equivalent to spin $\hbar = 1$ is $E_p = 1$, while the typical masses of particles are about $10^{-22}$.

IOP Conf. Series: Journal of Physics: Conf. Series **942** (2017) 012006 doi:10.1088/1742-6596/942/1/012006

## 2. Supersymmetric bag model compatible with spinning Kerr-Newman solution

The spinning Kerr-Newman (KN) solution [3, 12] is of particular interest, since, as it ws obtained by Carter, it has gyromagnetic ratio $g = 2$, the same as that of the Dirac electron [12, 13].

Metric of KN solution in the Kerr-Schild form is [12]

$$g_{\mu\nu} = \eta_{\mu\nu} + 2Hk_\mu k_\nu, \tag{2}$$

where $\eta_{\mu\nu}$ is metric of auxiliary Minkowski space $M^4$, (signature $(-+++)$), where scalar function

$$H_{KN} = \frac{mr - e^2/2}{r^2 + a^2\cos^2\theta}, \tag{3}$$

is given in oblate spheroidal coordinates $r$ and $\theta$, determined by transformations [12],

$$x + iy = (r + ia)\exp\{i\phi_K\}\sin\theta, \quad z = r\cos\theta, \quad \rho = r - t. \tag{4}$$

The null field $k_\mu(x)$, $(k_\mu k^\mu = 0)$ determines directions of 'frame-dragging', which takes fortex form Fig.1.,Fig.2. For ultra-high spin of spinning particles, $a = J/m >> m$, 'frame-dragging' becomes so strong that black hole horizons disappear and field $k^\mu(x)$ focuses on singular ring $r = 0, \cos\theta = 0$. The congruence $k^-_\mu$, which towards the disk $r = 0$ as 'in-going', intersects disk and turns into an 'out-going' one $k^+_\mu$, which is extended analytically on the same background, forming a second sheet of the metric. The space becomes two-sheeted, with two different metrics on the same background, $g^\pm_{\mu\nu} = \eta_{\mu\nu} + 2Hk^\pm_\mu k^\pm_\nu$. The Kerr singular ring $r = 0,\ \cos\theta = 0$ of the Compton radius $a = \hbar/2m$ breaks space topologically, playing the role of a branch line of the Kerr space.

Strange twosheetedness of Kerr space and remarkable coincidence with gyromagnetic ratio of the Dirac electron led to study of source of the Kerr solution as a classical model of electron mainly in two directions:

A) W. Israel [4] first suggested to truncate negative sheet of Kerr space along disk $r = 0$ and put at the disk a distribution of matter in accord with right hand side of the Einstein equations. He got a very strange matter distribution without a clear physical interpretation, however, he found an analogue with a quantum process of mass renormalization and first declared that this disk-like source of Compton radius can be considered as a classical model of electron. Very important consequent calculation by Hamity [14] showed that the matter distribution of the source took diagonal form in the corotating coordinate system, and thus, the source of Kerr solution turned into a relativistically rotating disk, boundary of which is moving with speed of light. Later on López generalized this model to model of a rotating bubble [6].

B) Another stringy version of the Kerr source was developed in our works with Prof. D. Ivanenko [5, 15], where Kerr singular ring was compared with Nielsen-Olesen (NO) vortex string model in superconductor [16]. It was shown that Kerr's singular ring can be considered as a lightlike string which can bear the electromagnetic and spinor traveling waves. Much later, such pp-wave strings were considered as solitonic string solutions to low energy string theory [17, 18].

In the both these directions of investigation source of the Kerr geometry created a gravitational vortex of the Compton size, and there appeared the question: Why this vortex was not observed experimentally, where electron looks like a point without structure? Answer follows from mechanism of Lorentz contraction [19] (see also metric on relativistically rotating disk in Landau and Lifshitz book The Classical Theory of Fields, Sec.89) – *for external observer Kerr's vortex, as well as the lightlike circular string of Compton radius shrink to point by Lorentz contraction.*

The suggested in [8, 9, 10] supersymmetric bag model merges the directions A) and B) on the base of Higgs mechanism and simultaneously solves the problem of unification with gravity.

IOP Conf. Series: Journal of Physics: Conf. Series **942** (2017) 012006 doi:10.1088/1742-6596/942/1/012006

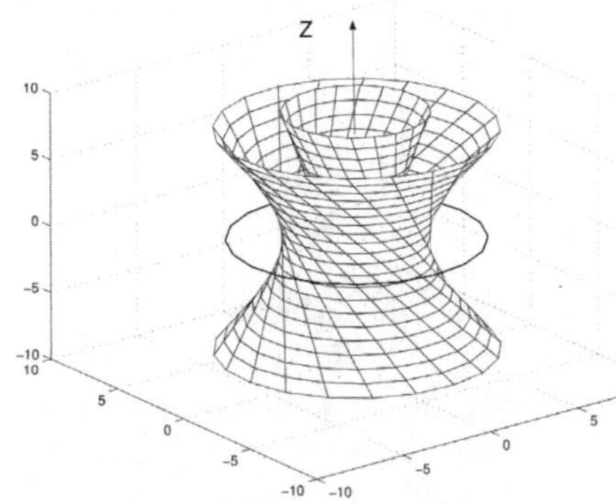


**Figure 1.** The lightlike Kerr Null congruence $k^\mu$ determines space-dragging caused by mass and spin.

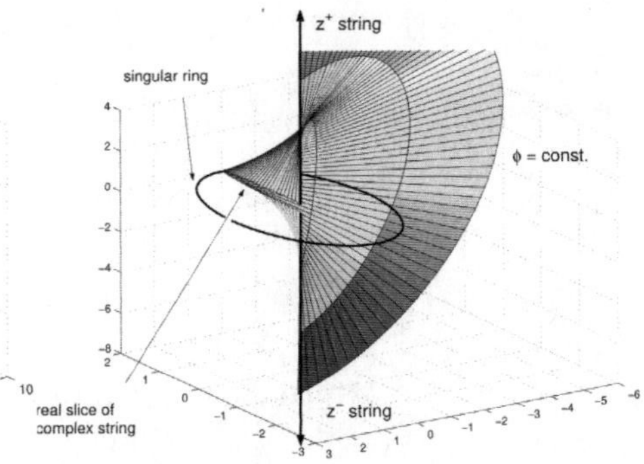


**Figure 2.** Rotational dragging of vector-potential $A^\mu$ along border of bag creates a quantum Wilson loop.

Quantum theory requires flat space, at least in the Compton zone, but spinning Kerr's gravity spoils space topologically, and it is unclear, how quantum theory can work on such background. Supersymmetry expels gravitational field from Compton zone of spinning particle, similar to expulsion of electromagnetic field from superconductor. Supersymmetric bag model realizes such expulsion of gravity and electromagnetic field, forming tree zones: flat quantum interior **(I)**, external zone with exact KN solution **(E)**, and a thin zone of transition **(R)**, see Fig.3.

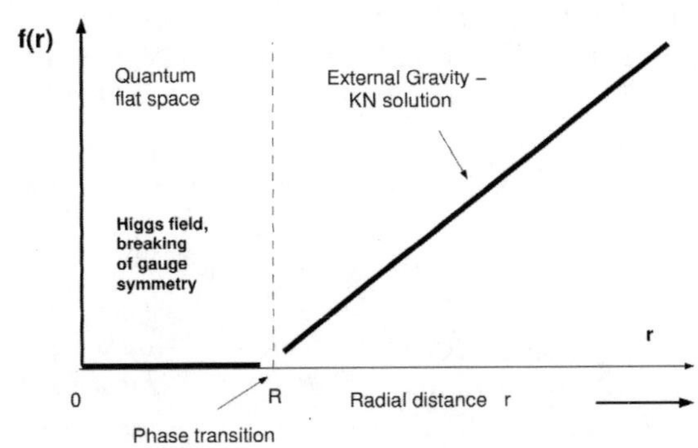


**Figure 3.** Separation of the flat Quantum interior **(I)** from zone of KN Gravity **(E)** by domain wall phase transition **(R)**.

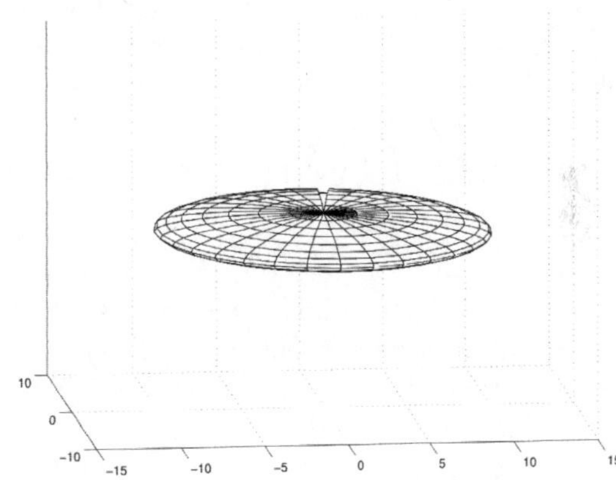

**Figure 4.** Shape of the KN bag.

For the giant values of spin, these demands become so restrictive that structure of bag is determined almost unambiguously.

Surface $(R)$ is defined by the continuous transition of KN solution to Minkowski interior of the bag, (C. López [6]). According (2) and (3), zone **(R)** corresponds to $H_{KN}(r) = 0$, which gives

$$r = R = e^2/2m, \tag{5}$$

and (4) shows that boundary of bag must form an ellipsoidal domain wall, the disk with radius $r_c = \sqrt{R^2 + a^2}$ and thickness $R$, see Fig.4.

To satisfy **(I)**,**(E)**,**(R)**, it is natural to use Higgs mechanism of symmetry breaking which is used in many nonperturbative electroweak nodels and in the MIT and SLAC bag models

IOP Conf. Series: Journal of Physics: Conf. Series **942** (2017) 012006 doi:10.1088/1742-6596/942/1/012006

[20, 21]. The corresponding Lagrangian is also known as Landau-Ginzburg (LG) field model for superconducting phase transitions and used in famous Nielsen-Olesen (NO) model of vortex string in superconductor, [16],

$$\mathcal{L}_{NO} = -\frac{1}{4}F_{\mu\nu}F^{\mu\nu} - \frac{1}{2}(\mathcal{D}_\mu\Phi)(\mathcal{D}^\mu\Phi)^* - V(|\Phi|), \tag{6}$$

where $\mathcal{D}_\mu = \nabla_\mu + ieA_\mu$ are $U(1)$ covariant derivatives, $F_{\mu\nu} = A_{\mu,\nu} - A_{\nu,\mu}$ the corresponding field strength. Usually, potential $V$, has the form $V = \lambda(\Phi^\dagger\Phi - \eta^2)^2$, which is inappropriate, since it breaks gauge symmetry in external zone (E), distorting gravitational field. To satisfy all the conditions **(I)**,**(E)**,**(R)** one should use the supersymmetric LG model with three Higgs-like fields, [22] $(H, Z, \Sigma) \equiv (\Phi_1, \Phi_2, \Phi_3)$.

Corresponding Lagrangian differs from (6) only by summation over the fields $\Phi_i, i = 1, 2, 3,$ while the potential $V$ should be formed from a superpotential $W(\Phi_i)$,

$$V(r) = \sum_i F_i F_i^*, \quad F_i = \partial W/\partial\Phi_i \equiv \partial_i W, \tag{7}$$

where

$$W(\Phi_i, \bar{\Phi}_i) = Z(\Sigma\bar{\Sigma} - \eta^2) + (Z + \mu)H\bar{H}, . \tag{8}$$

The conditions $F_i = \partial_i W = 0$ determine two vacuum states with $V = 0:$

internal- **(I)**, where $r < R - \delta$, and Higgs field $|H| = \eta$, and

external- **(E)**, where $r > R + \delta$ and Higgs field $H = 0$, separated by domain wall with $V > 0$, forming ellipsoidal boundary – zone **(R)**, in correspondence with **(I)**,**(E)**,**(R)**.

Note that quartic potential used in known bag models forms a "cavity in superconductor", while the supersymmetric potential (7)-(8) concentrates Higgs field in zone **(I)**, giving supersymmetric and superconducting vacuum state inside bag, where we can use the NO Lagrangian (6) leading to equations

$$\partial_\nu\partial^\nu A_\mu = J_\mu = e|H|^2(\chi,_\mu + eA_\mu), \tag{9}$$

where $\chi$ is oscillating phase of Higgs field $H = |H|e^{i\chi(t,\phi)}$. Vector potential of the KN solution concentrates in equatorial plane near the Kerr singular ring and forms a circular string which is similar to Nielsen-Olesen vortex string in superconductor, but placed on the boundary of superconducting disk at $r = R = e^2/2m$ where it taks maximal value $A^*_\mu dx^\mu = -Re\,\frac{2m}{e}(dt - ad\phi)$ and forms a longitudinal lightlike field dragged in $t, \phi$ direction by the Kerr congruence near the ring, see Fig. 4. It forms a stringy Wilson loop which is compensated by phase of Higgs field, $\chi,_\mu + eA_\mu = 0 \Rightarrow J_\mu = 0$ inside the disk, leading to $\chi = 2m(t + a\phi)$. Thus, current $J_\mu$ is expelled to surface layer of the superconducting disk. Periodicity of the phase $\chi(\phi) = \chi(\phi + 2\pi)$ leads to remarkable consequence – *quantization of angular momentum*, [8, 9, 10], $J = n/2, \quad n = 1, 2, 3, ....$

By nature bag models are assumed to be soft and flexible [25, 23, 24]. Under excitations bags are deformed and take the shape of a stringy flux-tube joining the quark-antiquark pair [20, 23], or toroidal string [21, 24, 25].

The bag adopted to KN gravitational field acquires an ellipsoidal shape, see Fig.4, and circular string is formed on the boundary of the disk, closely to Kerr singular ring. So, the string is really formed by singular ring and regularized by bag boundary. The assumption, that Kerr singular ring is similar to NO model of dual string was done long ago in [5, 15], where it was noted that excitations of the KN solution create traveling waves along the Kerr ring. Later, it was obtained in [17] close connection of the Kerr singular ring with the Sen fundamental string solution to low energy string theory, and other relations of Kerr geometry with string theory

IOP Conf. Series: Journal of Physics: Conf. Series **942** (2017) 012006 doi:10.1088/1742-6596/942/1/012006

(see refs. in [9]). String admits traveling waves, which deform position of the bag boundary according **(R)**, creating a circulating lightlike node, where surface of the deformed bag touches the Kerr singular ring, creating a circulating lightlike singular pole, which can be associated with a confined quark, Fig.4B, and super-bag turns into a single "bag-string-quark" system, resembling D2-D1-D0-brane sistem of string–Mtheory.

In turn, capture of quarks, is one more special feature of the bag model, requiring consistent implementation of the Dirac equation [20, 21, 24, 25]. In KN geometry, it is defined according to famous Kerr Theorem [12] which defines the shear-free Kerr congruences in twistor terms, and gives two roots $Y^{\pm}$ for projective spinor coordinate

$$Y = \phi_1/\phi_0, \tag{10}$$

which is equivalent to two-component Weyl spinor $\phi_\alpha$, and defines the null direction as

$$k_\mu = \bar{\phi}_{\dot{\alpha}} \sigma_\mu^{\dot{\alpha}\alpha} \phi_\alpha. \tag{11}$$

As it was shown in [26, 8], two roots of the Kerr theorem $Y^{\pm}$ give us two Weyl spinors, $\phi_\alpha$ and $\bar{\chi}^{\dot{\alpha}}$, consistent with ingoing and outgoing KN solutions in zone **(E)**. Inside the bag, these solutions unite, forming the Dirac spinor

$$\Psi = \begin{pmatrix} \phi_\alpha \\ \bar{\chi}^{\dot{\alpha}} \end{pmatrix} \tag{12}$$

which gets mass through Yukawa coupling to condensate of the Higgs field.

Finally, Super-Bag can be naturally upgraded to Wess-Zumino SuperQED model, [11], revealing connections between the non-perturbative solutions of the supersymmetric LG model and the conventional perturbative technics used in QED.

## 3. Conclusion

The giant spin/mass ratio of spinning particles shows that the known earlier concept of weakness of gravitational interaction was incorrect. The shift scale interaction from Planck to Compton distances opens new way to consistent with gravity particle physics based on *supersymmetric Higgs model* which resolves the conflict between these theories without quantization of gravity or modifications of the Einstein-Maxwell field equations.

Using supersymmetric Higgs (LG) field model we find a nonperturbative supersymmetric bag solution, which separates the external curved space from Compton zone of quantum processes. Bag model naturally merges two old directions of investigation of the problem of source of the KN solution, A) and B).

Applying this model to electron, we should note the fact that for electron $g_e \approx 2$ plus the correction $\alpha/\pi$. The KN model gives only $g_e = 2$ as the Dirac model before QED. Of course, the considered KN model is not final, since many effects are not taken into account. In particular, the effect of back reaction of the excitations cannot be considered at present time, since the accelerated KN solutions are not obtained so far because the extraordinary difficulty of the mathematical part of this problem.

Note also, that $r_K$ is just the Compton zone of the dressed electron which is responsible for the appearance of the Lamb shift which is described perturbatively in QED. The described superBag model is nonperturbative and there appears natural requirement to reproduce all remarkable results obtained by QED. The recent paper [11] is just the work in this direction, since it states a link between perturbative QED and remarkable cancellations between component diagrams obtained for the perturbative version of the supersymmetric Wess-Zumino superQED model.

There are a lot of other open problems in this direction of investigation.

IOP Conf. Series: Journal of Physics: Conf. Series **942** (2017) 012006 doi:10.1088/1742-6596/942/1/012006

## 4. References


[1] Becker K, Becker M and Schwarz J 2007 *String Theory and M-Theory: A Modern Introduction*
[2] Schwarz J, *The Early History of String Theory and Supersymmetry*, CALT-68-2858 [arXiv:1201.0981]
[3] Kerr R P 1963 *Phys.Rev.Letters* **11**, 237
[4] Israel W 1970 *Phys. Rev.* D **2** 641
[5] Burinskii A Ya 1974 *Sov. Phys. JETP* **39** 193
[6] López C A 1984 *Phys. Rev.* D **30** 313
[7] Misner Ch W, Thorne K S and Wheeler J A 1973 *Gravitation* v.3, San Francisco: W. Freeman
[8] Burinskii A 2015 *JETP (Zh. Eksp. Teor. Fiz.)* **148**(8) 228, [arXiv:1505.03439]
[9] Burinskii A 2015 *JETP (Zh. Eksp. Teor. Fiz.)* **148** (11) 937, [arXiv:1706.02979]
[10] Burinskii A 2016 *Phys Lett* B **754** 99
[11] Burinskii A 2017, arXiv:1701.01025
[12] Debney G C, Kerr R P and Schild A 1969 *J. Math. Phys.* **10** 1842
[13] Carter B 1968 *Phys. Rev.* **174**, 1559
[14] Hamity V 1976 *Phys. Lett.* A **56**, 77
[15] Ivanenko D D and Burinskii A Ya 1975 *Izv. Vuz. Fiz.*, **5**, 135
[16] Nielsen H B and Olesen P 1973 *Nucl. Phys.* B **61** 45
[17] Burinskii A 1995 *Phys. Rev.* D **52** 5826 [arXiv:hep-th/9504139]
[18] Dabholkar A et. al. 1996 *Nucl Phys* **B474** 85
[19] Punsly B 1985 *J. Math. Phys.* **26** 1728
[20] Chodos A. et al. 1974 *Phys. Rev.* D **9** 3471
[21] Bardeen W. A. at al. 1974 *Phys. Rev.* D **11** 1094
[22] Morris J R 1996 *Phys. Rev.* **D 53** 2078, [arXiv:hep-ph/9511293]
[23] Johnson K and Thorn C B 1976 *Phys. Rev.* **D 13** 1934
[24] Tye S-H H 1976 *Phys. Rev.* **D 13**, 3416
[25] Giles R C and Tye S-H H 1976 *Phys. Rev.* **D 13**, 1690
[26] Burinskii A 2015 *Grav. and Cosmol.* **21**(1) 28, [arXiv:1404.5947].

IOP Conf. Series: Journal of Physics: Conf. Series **942** (2017) 012007 doi:10.1088/1742-6596/942/1/012007

# Observational signatures of spherically-symmetric black hole spacetimes

**Mariafelicia De Laurentis, Ziri Younsi, Oliver Porth, Yosuke Mizuno, Christian Fromm, Luciano Rezzolla, Hector Olivares**

Institut für Theoretische Physik, Max-von-Laue-Straße 1, D-60438 Frankfurt am Main, Germany

E-mail: laurentis@th.physik.uni-frankfurt.de

**Abstract.** A binary system composed of a supermassive black hole and a pulsar orbiting around it is studied. The motivation for this study arises from the fact that pulsar timing observations have proven to be a powerful tool in identifying physical features of the orbiting companion. In this study, taking into account a general spherically-symmetric metric, we present analytic calculations of the geodesic motion, and the possible deviations with respect to the standard Schwarzschild case of General Relativity. In particular, the advance at periastron is studied with the aim of identifying corrections to General Relativity. A discussion of the motion of a pulsar very close the supermassive central black hole in our Galaxy (Sgr A*) is reported.

## 1. Introduction

It is now widely believed that supermassive black holes (SMBHs) reside at the centres of all galaxies and that their estimated masses are in the range of a few million to up to tens of billions of solar masses. Earth's closest SMBH candidate is found at the Galactic Centre, Sagittarius A* (Sgr A*), which astronomers have been observing for several decades [1]. It is expected that the mathematical description of astrophysical black holes (BHs) is based on solutions to the Einstein field equations, and therefore founded on general relativity (GR). However, there also exist many BH solutions in extended and alternative theories of gravity, and to-date observational constraints, most notably in the strong-field regime, are lacking [2]. One promising probe of strong-field gravity is the direct imaging of the shadow cast by a SMBH[3]. Direct observation of the event horizon of our own SMBH, Sgr A*, will soon be obtained by the Event Horizon Telescope Collaboration (EHTC)[4, 5]. Pulsars provide an additional independent observational tool to help improve the understanding of the properties of Sgr A* (e.g., its mass, spin and even geometry), providing considerably stronger constraints than is possible with event horizon-scale imaging alone. The case of a pulsar orbiting around a SMBH is particularly interesting since one can in principle combine precision timing measurements with measurements of geodesic motion around the BH, i.e., in the strong-field regime. Such timing measurements can contribute to fixing strict ranges on the parameters of a given class of gravity theories and therefore facilitate the selection of viable theories without imposing any *a priori* assumptions. However, in this study theories of gravity entirely distinct from GR are investigated, and consequently the parametrisation of Rezzolla and Zhidenko (RZ) [6] is employed. The RZ parametrization is a general representation of BH spacetimes in arbitrary metric theories of gravity. In the case of spherically symmetric spacetimes, the parametrization makes use of a coordinate compactification in terms of a rapidly-convergent continued-fraction expansion defined in the radial direction between the event horizon and spatial infinity. The focus of the present

IOP Conf. Series: Journal of Physics: Conf. Series **942** (2017) 012007 doi:10.1088/1742-6596/942/1/012007

study is restricted to the spherically symmetric case and general expressions for the dynamics of a test-particle in general BH spacetimes, such as the motion of a pulsar orbiting around the SMBH candidate Sgr A*, are derived. In particular, explicit general expressions for the advance of the periastron at zeroth order of the parametrization is provided. Furthermore, periastron-advance formulae is also given for Einstein-Maxwell-Axion-Dilaton theory [7].

## 2. Parametrization Framework

In what follows the RZ parametrization [6] for a generic spherically symmetric BH spacetime is briefly reviewed and subsequently used to determine the dynamics of a test particles. The Lagrangian may be written as [7]

$$2\mathcal{L} = N(r)\dot{t}^2 - \frac{B(r)}{N(r)}\dot{r}^2 - r^2\dot{\theta}^2 - r^2\sin^2\theta\dot{\varphi}^2 , \tag{1}$$

In the RZ parametrization, the function $N(r)$ is then expressed as $N(x) = xA(x)$ where $A(x) > 0$ for $0 \le x \le 1$, with $x := 1 - \frac{r_0}{r}$, so that $x = 0$ is the position of the event horizon and $x = 1$ corresponds to spatial infinity. Furthermore, $A$ and $B$ may be expressed in terms of the parameters $\epsilon$, $a_i$, and $b_i$ ($i \in [0, n]$, where $n$ is the expansion order), such that

$$A(x) = 1 - \epsilon(1-x) + (a_0 - \epsilon)(1-x)^2 + \widetilde{A}(x)(1-x)^3 , \tag{2}$$

$$B(x) = 1 + b_0(1-x) + \widetilde{B}(x)(1-x)^2 , \tag{3}$$

where the functions $\widetilde{A}$ and $\widetilde{B}$ describe the metric near the horizon (i.e., $x \simeq 0$) and at spatial infinity (i.e., $x \simeq 1$). It is evident that the metric is finite in both limits [6]. The functions (2)–(3) can then be expanded via a Padé approximation of continuous as

$$\widetilde{A}(x) = \cfrac{a_1}{1 + \cfrac{a_2 x}{1 + \cfrac{a_3 x}{1 + \ldots}}} , \quad \widetilde{B}(x) = \cfrac{b_1}{1 + \cfrac{b_2 x}{1 + \cfrac{b_3 x}{1 + \ldots}}} , \tag{4}$$

where $a_1, a_2, \ldots, a_n$ and $b_1, b_2, \ldots, b_n$ are dimensionless constants that can be fixed once the generalised metric is matched to a specific metric. Finally, the parameter $\epsilon$ in equations (2)–(3) measures the deviations of the position of the event horizon in the general metric from the corresponding location in a Schwarzschild spacetime, i.e., $\epsilon = \frac{2M - r_0}{r_0} = -\left(1 - \frac{2M}{r_0}\right)$.

## 3. Applications of the RZ parametrization: Einstein-Maxwell-Axion-Dilaton metric

The Einstein-Maxwell-Axion-Dilaton (EMAD) metric considered in this study is spherically symmetric and is constructed from a simplification of the axisymmetric EMAD solution [8] in the case of a vanishing axion field. Solutions of this type arise from string theory. When the axion field vanishes and the BH is spherically symmetric, the EMAD BH ine element takes the following form

$$ds^2 = 1 - \frac{2M\rho}{r^2}dt^2 - \frac{r^2}{\hat{b}^2 + r^2}dr^2 - d\Omega^2 \tag{5}$$

where $\mu := M - \hat{b}$. Here $\hat{b}$ is the dilaton parameter and M the BH mass [9], and $\rho^2 = r^2 + 2\hat{b}r$, where $r \equiv r(\rho)$. Recalling that the location of the event horizon for this BH is given by $\rho_{0,\mathrm{EMAD}} = 2\left(M - \hat{b}\right)$,

The expression for $\epsilon$ in terms of the axion-dilaton parameters may be written as $\epsilon_{\mathrm{EMAD}} = \sqrt{1 + \frac{\hat{b}}{\mu}} - 1$, In a similar manner, expanding the metric coefficients at spatial infinity gives the values for $a_0$ and $b_0$ as

$$a_{0,\mathrm{DIL}} = \frac{\hat{b}}{2\mu} , \quad \text{and} \quad b_{0,\mathrm{DIL}} = 0 . \tag{6}$$

IOP Conf. Series: Journal of Physics: Conf. Series **942** (2017) 012007 doi:10.1088/1742-6596/942/1/012007

**Table 1.** Values of the periastron advance $\Delta\varphi := 2\pi\,\Delta\widetilde{\varphi}$ for different objects. Here the numerical values for GR displacement, $\Delta\varphi_{\rm GR}$ and for the RZ parametrization (7) are shown. The table reports the measured values of the eccentricity $e$, semi-major axis $a$, and $\sigma = r_{\rm g}/\ell$, assuming that the gravitational radius is $r_{\rm g} \sim 6.64657 \times 10^9{\rm m} = 0.044\,{\rm AU}$ for Sgr A*, and that the values of coefficients $a_0$and $\epsilon$ are all set to $10^{-3}$. Numbers in square brackets denote multiplicative powers of ten.

| Object | $e$ | $a\,[{\rm AU}]$ | $\sigma$ | $\Delta\varphi_{\rm GR}$ | ${}^{(0)}\Delta\varphi$ |
|---|---|---|---|---|---|
| Toy I | 0.800 | 1.75400[+2] | 7.03608[−4] | 1.32627[−2] | 1.33025[−2] |
| Toy II | 0.800 | 4.38500[+1] | 2.81443[−3] | 5.30508[−2] | 5.32097[−2] |
| Toy III | 0.786 | 5.00000[+0] | 2.32488[−2] | 4.38229[−1] | 4.39524[−1] |
| Toy IV | 0.888 | 1.00000[+0] | 2.10110[−1] | 3.96047[+0] | 3.97069[+0] |

As a direct application of the framework developed in the previous sections, the transition between two close inner orbital turning points (or, equivalently, between two close outer turning points) is calculated. The orbits in this case remain closed if the magnitude $\Delta\varphi$ of the angle swept out by the orbit is $2\pi$. If this is not the case, then the inner turning points are precessing and the amount of this precession per orbit is $\delta\varphi_{\rm prec} = \Delta\varphi - 2\pi$ . To illustrate how to derive a theory-independent expression for the periastron advance within the RZ parametrization, first consider the expansion of the metric when the only non-zero expansion terms are $a_0$ and $b_0$. The expression for the periastron advance, $\Delta\varphi$, at zeroth order in the RZ parametrization is obtained as

$$ {}^{(0)}\Delta\widetilde{\varphi} = 3\sigma + \sigma\left(9 - 26\sigma + 20\sigma^2\right) a_0 + 20\sigma^2\left(1-\sigma\right)\epsilon\,, \tag{7} $$

where $\Delta\widetilde{\varphi} := \Delta\varphi/2\pi$ and $\sigma = r_{\rm g}/\ell$ with $\ell$ is the semi-latus rectum of the orbits. The first term in the above equation is simply the GR contribution, while the remaining two terms represent the deviations from GR in a general BH spacetime at zeroth order in the RZ parametrization.

## 4. Application to astrophysical test cases

In the this sections both the RZ parametrization and the expression derived for the periastron advance are tested. To do this, a set of four test objects (four pulsar toy models) are considered. Their properties have been chosen to have a range of semi-major axes, reasonably high eccentricities and moderate-to-short orbital periods. In this respect, these toy models are idealised, but not altogether unrealistic: future advances in instrumental sensitivity could, in principle, enable the detection of Galactic Centre pulsars with such properties [12]. In modelling the toy pulsar-SMBH system it is hereafter assumed that the mass of the central SMBH is known to some degree of precision.

The properties of these pulsars are collected in Table 1, which also reports, besides the eccentricities $e$ and semi-major axes $a$, the values of the periastron advances $\Delta\widetilde{\varphi}_{\rm GR}$ and ${}^{(0)}\Delta\widetilde{\varphi}$ at the zeroth order, in the RZ expansion. The models I, II, III are in principle already measurable with present radio-astronomical observations [13], while Toy IV is, at the present time, an optimistic model. Since the RZ parametrisation is constructed to be most accurate at the event horizon and at spatial infinity, the results presented in Table 1 are, albeit weakly, dependent on the specific form of the parametrization. Note that in evaluating the periastron advance, specific values for the coefficients $\epsilon, a_i$ and $b_i$ must be specified, since these coefficients cannot yet be constrained by astrophysical observations and hereafter chosen to be $a_0 = \epsilon = 10^{-3}$. In particular, looking at Table 1 one can establish how well the RZ parametrization works in the vicinity of the event horizon (this is especially true for Toy IV). This is illustrated in Fig. 1 for the EMAD BH (assuming $\hat{b} = 5 \times 10^{-8}$) using models Toy II and Toy IV, where the periastron advance (light-blue shaded areas around the solid blue line) and of the orbital period (light-red shaded areas around the red line) assuming an accuracy of $10^{-4}$. In summary, the results reported in Figs. 1 demonstrate that by using a general description of test-particle motion in arbitrary BH spacetimes,

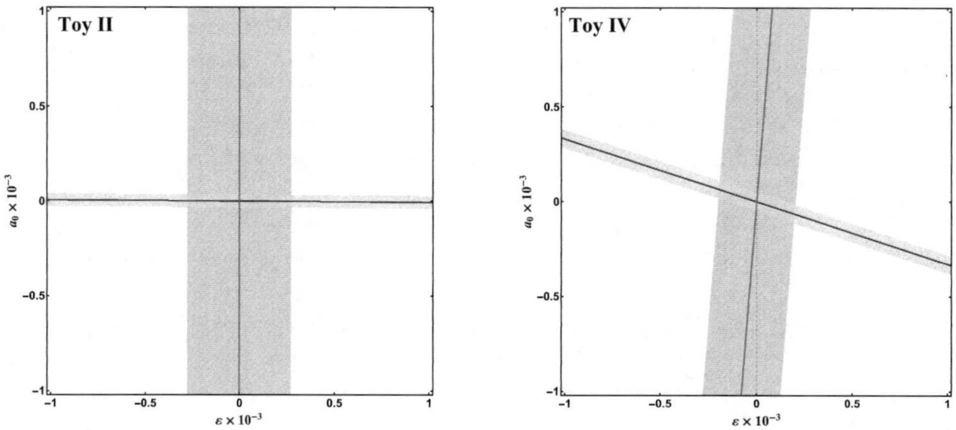


**Figure 1.** Constraints set by the pulsar orbits for toy models II and IV in the EMAD metric (5) at zeroth order, with dilaton parameter fixed at $\hat{b} = 5 \times 10^{-8}$. The values of the coefficients $a_0$ and $\epsilon$ are constrained by the observations of the periastron advance (light-blue shaded areas around the solid blue line) and of the orbital period (light-red shaded areas around the solid red line) assuming an accuracy of $10^{-4}$.

such as the RZ parameterization, future observations of pulsars near a SMBH can help impose tighter constraints on different theories of gravity, and even potentially facilitate ruling out certain theories (and, by extension, related classes and extensions thereof) entirely.

Furthermore, an illustrative example demonstrating the facility of the parametric framework to constrain the parameters of a theory of gravity, is presented. Figures 2 present the deviations from GR of the periastron advance as a function of the semi-major axis length, for models Toy I, Toy II (left panels) and model Toy IV (right panels) (Fig. 2). Multiple coloured curves, coloured from violet through to red, denote, for a fixed value of the theory parameter, how the relative difference in the periastron advance varies as a function of the semi-major axis length. These 51 coloured lines are uniformly logarithmically spaced between the stated parameter value in the upper right of each panel (uppermost red line) and 0.01 times that value (bottommost violet line), i.e., 25 lines per decade in the theory parameter. The horizontal black dashed line at $10^{-7}$ represents a potential astrophysical measurement precision [13]. It is immediately clear that for models Toy I and Toy II the relative differences in the periastron advance are insensitive to the semi-major axis length and practically indistinguishable, as is evident from the horizontal, parallel theory parameter lines. This can be interpreted as near-Keplerian pulsar motion, i.e., the weak-field limit of the RZ parametrization. The vertical blue line (right panel) denotes the semi-major axis position of model Toy IV. Since this model places the pulsar much closer to the event horizon of the BH (1 AU), and as is clear from the theory parameter lines steeper gradients, the pulsar motion can be considered as occurring in the transition region between the weak-field and strong-field regimes. The intersection of the uppermost plotted theory parameter line with the vertical blue and magenta lines provides an upper limit for the theory parameters $\hat{b} < 6.68 \times 10^{-8}$ for Toy I and Toy II, whereas for Toy IV case this yields $\hat{b} < 7.762 \times 10^{-8}$. In particular, one can see from Fig. 2 that it is possible to provide constraints on the two theories. More specifically, one can place an upper limit on the value of the parameters $\alpha$ and $\hat{b}$. It is assumed that the error on the measurements is of the order of $10^{-7}$ [13]. Figure 2 present the relative differences from ${}^{(0)}\Delta\widetilde{\varphi}$ at first order of the expansion in GR with respect to the semi-major axes for the four Toy models. What emerges from the figures is that the deviations from GR are much more remarkable if one considers, as seen before, small semi-major axes and high eccentricities. In addition, one may also estimate the various values of the parameters that characterise

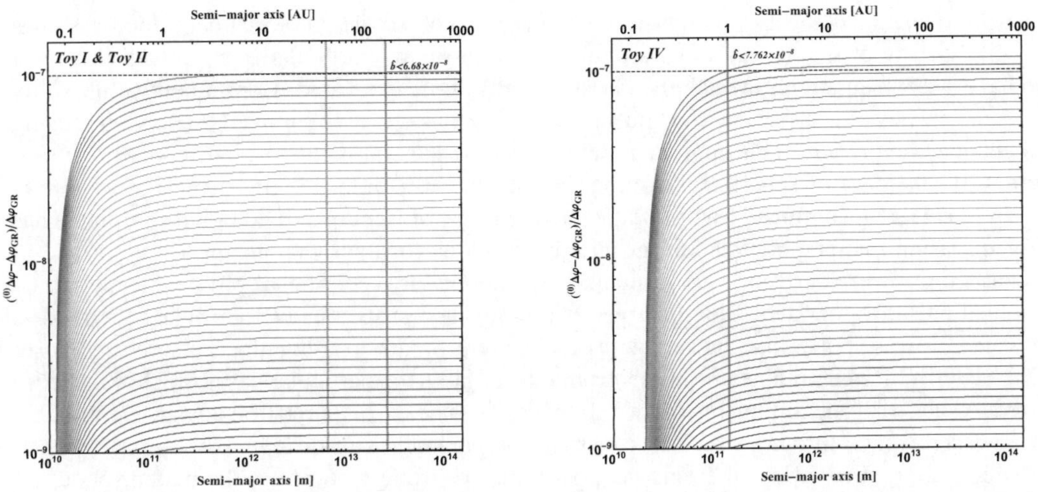


**Figure 2.** Relative difference of the zeroth order periastron advance with respect to the GR value plotted as a function of semi-major axis distance for Toy models I,II (left panel) and IV (right panel). Different coloured lines for a fixed value of the theory parameter the variation in this relative difference as a function of semi-major axis.The 51 lines are equally spaced (logarithmically) between the stated inset upper limits of $\hat{b} \leq 6.68 \times 10^{-8}$ (left panel) and $\hat{b} \leq 7.762 \times 10^{-8}$ (right panel). In the left panel solid lines at 175.4 AU (blue) and 43.85 AU (magenta), and in the right panel (blue) at 1 AU denotes the semi-major axis position of Toy model I, II and IV respectively.

the specific theory through the location of the different level curves. It can be seen from Figs. 2 that for a given measurement precision, as the pulsar semi-major axis location is shifted towards the BH event horizon, the range over which the theory parameter can be probed (and effectively constrained) is increased. Therefore, a pulsar orbiting in the immediate vicinity of a BH event horizon (i.e., in the truly strong-field regime) can provide much tighter constraints on the theory parameters. It is clear that in order to be proven useful, such a pulsar must be close to the SMBH, with an orbital period of only a few hours or less. Observational detection of such pulsars is in principle achievable with present day radio telescopes.

## 5. Conclusions

Although GR has proven to be a reliable theory of gravity in several different fields of application, and across several different scales, it is by no means a complete one. Furthermore, there exist other theories of gravity which reproduce not only the results of classical GR, but also the results of modern astrophysical observations where pure GR fails to be as predictive. There is therefore a pressing need, both theoretically and observationally, to begin to, at the very least, impose constraints (and even potentially exclude) particular theories of gravity. Owing to their very narrow mass range, extreme compactness and rapid, stable rotation periods, pulsars are one of the best candidates to probe strong-field gravity in the truly nonlinear regime. Measurement of a pulsar in the strong-field regime, i.e., near a BH event horizon, would enable not just a highly accurate determination of the BH properties, but also provide an accurate probe of the spacetime structure and geometry. Given the rapid increase in recent efforts to perform astronomical observations of the Galactic Center, the prospect of detecting a pulsar orbiting in close proximity to Sgr A* is promising. Such a detection would provide the most accurate measurements of the physical parameters (e.g. mass, spin, and even quadrupole moment) of Sgr A* [14]. Theoretical studies of pulsar motion and timing in both GR and alternative theories of gravity are therefore of great importance. However, given the breadth of available theories of gravity

IOP Conf. Series: Journal of Physics: Conf. Series **942** (2017) 012007 doi:10.1088/1742-6596/942/1/012007

in the present literature, it is most expedient to perform any such studies in a manner which assumes neither a specific theory of gravity nor any particular solution to any particular theory. It is also most desirable to have a representation in which the classical GR limit is recovered. Consequently, this study has presented an analysis of test-particle (i.e., pulsar) dynamics in several different BH spacetimes, using a theory-independent approach. This approach makes use of a general mathematical representation of BH spacetimes. In the case of spherically symmetric spacetimes, i.e., those considered in the present study, a rapidly-convergent continued fraction expansion in terms of a compactified radial coordinate has been employed. Using this parametrization, general expressions for the dynamics of a test-particle in general BH spacetimes were derived. This formalism was applied to a set of astrophysical test case: (i) four hypothetical pulsar toy models. Next, the periastron advance properties of four pulsar toy models were investigated. It was shown that, pulsars with smaller semi-major axis lengths (i.e. orbiting closer to Sgr A*) indeed exhibit quantifiable deviations from GR. Finally, the particular cases of EMAD theory was employed to show that measurements of the relative difference of the periastron advance (from GR) can provide another avenue through which to constrain the parameters of different theories of gravity. Therefore, using pulsar observations presents the possibility to strongly constrain the parameters of all theories which are purely geometrical (i.e., not containing, e.g., exotic particles and scalar fields).

In conclusion, pulsar observations can in principle accurately constrain the properties of their central SMBH companion. In this study, using a theory-independent parametric framework, it has been shown that pulsar timing and dynamics also presents the possibility to constrain (and even potentially exclude) theories of gravity [7].

## Acknowledgements

Support comes from the ERC Synergy Grant "BlackHoleCam - Imaging the Event Horizon of Black Holes" (Grant 610058)

## References

[1] Eckart A and Genzel R 1996 *Nature* **383** 415–417
[2] Capozziello S and de Laurentis M 2011 *Phys. Rep.* **509** 167–321 (*Preprint* 1108.6266)
[3] Younsi Z, Zhidenko A, Rezzolla L, Konoplya R and Mizuno Y 2016 *Phys. Rev. D* **94** 084025 (*Preprint* 1607.05767)
[4] Doeleman S S, Weintroub J, Rogers A E E, Plambeck R, Freund R, Tilanus R P J, Friberg P, Ziurys L M, Moran J M, Corey B, Young K H, Smythe D L, Titus M, Marrone D P, Cappallo R J and et al 2008 *Nature* **455** 78–80 (*Preprint* 0809.2442)
[5] Goddi C, Falcke H, Kramer M, Rezzolla L, Brinkerink C, Bronzwaer T, Davelaar J R J, Deane R, de Laurentis M, Desvignes G, Eatough R P, Eisenhauer F, Fraga-Encinas R, Fromm C M, Gillessen S and et al 2017 *International Journal of Modern Physics D* **26** 1730001-239 (*Preprint* 1606.08879)
[6] Rezzolla L and Zhidenko A 2014 *Phys. Rev. D* **90** 084009 (*Preprint* 1407.3086)
[7] De Laurentis M, Younsi Z, Porth O, Mizuno Y and Rezzolla L 2017 (*Preprint* 1712.00265)
[8] García A, Galtsov D and Kechkin O 1995 *Phys. Rev. Lett.* **74** 1276–1279
[9] Sen A 1992 *Phys. Rev. Lett.* **69** 1006–1009 (*Preprint* hep-th/9204046)
[10] Ghez A M, Klein B L, Morris M and Becklin E E 1998 *Astrophys. J.* **509** 678–686 (*Preprint* astro-ph/9807210)
[11] Ghez A M, Salim S, Weinberg N N, Lu J R, Do T, Dunn J K, Matthews K, Morris M R, Yelda S, Becklin E E, Kremenek T, Milosavljevic M and Naiman J 2008 *Astrophys. J.* **689** 1044–1062 (*Preprint* 0808.2870)
[12] Lorimer D R and Kramer M 2012 *Handbook of Pulsar Astronomy*
[13] Liu K, Wex N, Kramer M, Cordes J M and Lazio T J W 2012 *Astrophys. J.* **747** 1 (*Preprint* 1112.2151)
[14] Psaltis D, Wex N and Kramer M 2016 *Astrophys. J.* **818** 121 (*Preprint* 1510.00394)

IOP Conf. Series: Journal of Physics: Conf. Series **942** (2017) 012008 doi:10.1088/1742-6596/942/1/012008

# Self-consistent geodesic equation and quantum tunneling from charged AdS black holes

**Gao-Ming Deng**[1,2]

[1] Institute of theoretical physics, Beijing University of Technology, Beijing 100124, China
[2] School of physics and space science, China West Normal University, Nanchong 637002, China

E-mail: gmd2014@emails.bjut.edu.cn

**Abstract.** Some urgent shortcomings in previous derivations of geodesic equations are remedied in this paper. In contrast to the unnatural and awkward treatment in previous works, here we derive the geodesic equations of massive and massless particles in a unified and self-consistent manner. Furthermore, we extend to investigate the Hawking radiation via tunneling from charged black holes in the context of AdS spacetime. Of special interest, the application of the first law of black hole thermodynamics in tunneling integration manifestly simplifies the calculation.

## 1. Introduction

It is widely believed that Hawking radiation [1,2] via tunneling could be advanced as a promising tool to have a closer insight into puzzling black holes. Nowadays, research on Hawking radiation has captured popular attention [3–8]. It is worth mentioning that Parikh-Wilczek's semi-classical tunneling method [9] has always prevailed in further investigating tunneling radiation from various black holes [10, 11]. With respect to Parikh-Wilczek's semi-classical tunneling picture, the derivation of particles' geodesic equations plays an important role in the study of black hole radiation. However, there existed some notable shortcomings in previous derivations. For the case of massive particles, the geodesics were awkwardly derived by employing the relation $v_p = v_g/2$ between the group velocity and the phase velocity, which is inconsistent with the variational principle of action. According to the variation principle, in General Relativity, geodesic equations should be defined by applying the variational principle on the Lagrangian action. What's more, the above treatment is not applicable for defining the geodesics of massless particles. Geodesic equations of massless particles used to be derived by setting $ds^2 = 0$ instead. Even worse, previous derivation involved inconsistent foundation, i.e. mixing together relativistic and non relativistic descriptions. Motivated by these facts, this paper concentrates on remedying the urgent shortcomings and extending to investigate the Hawking radiation via tunneling from charged black holes in anti-de Sitter (AdS) spacetime.

## 2. Charged AdS black holes and the Painlevé-like form

The spherical Reissner-Nordström AdS black hole in (3+1)-dimensional spacetime is usually described by the line element and the U(1) gauged potential

$$ds^2 = -\Delta dt_R^2 + \frac{dr^2}{\Delta} + r^2 \left(d\theta^2 + \sin^2\theta d\phi^2\right) , \tag{1}$$

IOP Conf. Series: Journal of Physics: Conf. Series **942** (2017) 012008 doi:10.1088/1742-6596/942/1/012008

$$F = d\mathcal{A}\,, \quad \mathcal{A} = -\frac{Q}{r}dt\,, \tag{2}$$

with

$$\Delta = 1 - \frac{2M}{r} + \frac{Q^2}{r^2} - \frac{\Lambda}{3}r^2\,, \tag{3}$$

where the parameters $M$ and $Q$ represent the ADM mass and the total charge of the black hole respectively, the cosmological constant $\Lambda$ takes a negative value, and $t_R$ is the time coordinate. The event horizon of the black hole is located at $r = r_+$, where $r_+$ is determined by the largest root of the equation $\Delta(r) = 0$.

The key trick to apply Parikh-Wilczek's semi-classical tunneling analysis is to introduce a coordinate system which is well-behaved at the horizon. Therefore, to remove the coordinate singularity in the metric (1), we perform a generalized Painlevé coordinate transformation

$$dt_R = dt + \frac{\sqrt{1-\Delta}}{\Delta}dr\,, \tag{4}$$

and obtain the following Painlevé-like line element

$$ds^2 = -\Delta dt^2 + 2\sqrt{1-\Delta}dtdr + dr^2 + r^2 d\theta^2 + r^2\sin^2\theta d\phi^2\,. \tag{5}$$

Obviously, the new form (5) is regular at the horizon and displays the stationary, nonstatic nature of the spacetime. Besides, more importantly, it satisfies the Landau's condition of the coordinate clock synchronization [12,13]. It should be emphasized that these attractive features are very advantageous for us to explore the tunneling process of the charged massive particles across the horizon.

## 3. Geodesic equations of massive and massless tunneling particles

As mentioned above, the derivation of geodesics is an important aspect in Parikh-Wilczek's semi-classical tunneling picture. However, in previous works, the geodesic equations were unnaturally and awkwardly derived. In this section, our attentions are focused on remedying the urgent shortcomings and attempting to define the geodesic equations of massive and massless particles in a unified and self-consistent way.

Generally speaking, both the null and timelike geodesics can be derived from the Lagrangian action with the help of the variational principle. Associated with a black hole solution, there often exists three conserved integration constants, namely, the energy $\mathcal{E}$ and angular momentum $L$ which correspond to the Killing vector $\partial_t, \partial_\phi$ respectively, as well as the Hamiltonian $\mathcal{H}$ that can be restricted by the 4-velocity normalization condition $\mathcal{H} = -mK/2$, $(K = 0, 1)$. Particularly, $K = 0, 1$ correspond to the cases of null and timelike geodesics. With these integration constants in hand, it's enough to determine the equations of motion for massive particles. Of special interest, the massless particles' geodesics can be derived by setting $K = 0$ or the rest mass $m = 0$ from the geodesic equations of massive particles. As a consequence, we can write out the geodesic equations of massive and massless particles in a unified and self-consistent way. In what follows, we first utilize this attractive improvement to derive the geodesics of massive and massless particles tunneling across the Reissner-Nordström black hole in AdS spacetime.

Working with the metric (5) and considering a charged massive particle, the corresponding Lagrangian [14] is given by

$$\begin{aligned}\mathcal{L} &= \frac{m}{2}g_{\mu\nu}\dot{x}^\mu\dot{x}^\nu + q\mathcal{A}_\mu\dot{x}^\mu \\ &= \frac{m}{2}\left(-\Delta\dot{t}^2 + 2\sqrt{1-\Delta}\dot{t}\dot{r} + \dot{r}^2 + r^2\dot{\theta}^2 + r^2\sin^2\theta\dot{\phi}^2\right) - \frac{qQ}{r}\dot{t}\,,\end{aligned} \tag{6}$$

IOP Conf. Series: Journal of Physics: Conf. Series **942** (2017) 012008 doi:10.1088/1742-6596/942/1/012008

in which $m, q$ are the rest mass and electric charge of the particle. Subsequently, the generalized momentum $P_\alpha = \partial\mathcal{L}/\partial\dot{x}^\alpha$ can be obtained,

$$P_t = m\left(-\Delta\dot{t} + \sqrt{1-\Delta}\dot{r}\right) - \frac{qQ}{r}\,, \tag{7}$$

$$P_r = m\left(\sqrt{1-\Delta}\dot{t} + \dot{r}\right)\,, \tag{8}$$

$$P_\theta = m\, r^2\dot{\theta}\,, \tag{9}$$

$$P_\phi = m\, r^2 \sin^2\theta\dot{\phi}\,. \tag{10}$$

Due to the fact that

$$\frac{dP_t}{d\tau} = \frac{\partial\mathcal{L}}{\partial t} = 0\,, \quad \frac{dP_\phi}{d\tau} = \frac{\partial\mathcal{L}}{\partial\phi} = 0\,, \tag{11}$$

where $\tau$ is the affine parameter, apparently, $t$ and $\phi$ act as the cyclic coordinates which correspond to the conserved generalized momenta $P_t$ and $P_\phi$, i.e.

$$P_t = \mathcal{E} = \text{Const}\,, \quad P_\phi = L = \text{Const}\,. \tag{12}$$

Moreover, with the help of the Legendre transformation, we get the Hamiltonian as follows

$$\begin{aligned}\mathcal{H} &= \dot{t}P_t + \dot{r}P_r + \dot{\theta}P_\theta + \dot{\phi}P_\phi - \mathcal{L}\\ &= \frac{m}{2}\left(-\Delta\dot{t}^2 + 2\sqrt{1-\Delta}\dot{t}\dot{r} + \dot{r}^2 + r^2\dot{\theta}^2 + r^2\sin^2\theta\dot{\phi}^2\right)\\ &= \frac{m}{2}\, g_{\mu\nu}\dot{x}^\mu\dot{x}^\nu\,.\end{aligned} \tag{13}$$

Considering the 4-velocity normalization condition

$$-\frac{2}{m}\mathcal{H} = K = \begin{cases} 0 & \text{the massless particle's case} \\ 1 & \text{the massive particle's case}\end{cases} \tag{14}$$

and combining with Eq. (12), $\dot{r}, \dot{t}$ can be worked out. It should be pointed out that the definition for the massless particle's geodesics is equivalent to its previous derivation by means of $ds^2 = 0$. In consequence, the geodesic equation can be calculated as

$$\begin{aligned}\bar{r} &= \frac{dr}{dt} = \frac{\dot{r}}{\dot{t}}\\ &= \Delta\left\{\sqrt{1-\Delta} \pm \frac{(E+qQ/r)}{\sqrt{(E+qQ/r)^2 - \Delta Y}}\right\}^{-1}\,,\end{aligned} \tag{15}$$

in which

$$Y = \frac{L^2}{r^2\sin^2\theta} + m^2\left(K + r^2\dot{\theta}^2\right)\,, \tag{16}$$

and the signs "$\pm$" correspond to the geodesics of ingoing and outgoing particles from the event horizon. To better understand the geodesic equation (15), without loss of generality, we take the geodesics of the plane $\theta = \pi/2$ as an example.

**i)** If setting $K = 1$ and considering the $s$-wave approximation ($L = 0$), it's easy to get the massive particle's geodesics equation in the plane $\theta = \pi/2$ as

$$\bar{r} = \Delta\left[\sqrt{1-\Delta} \pm \frac{(E+qQ/r)}{\sqrt{(E+qQ/r)^2 - m^2\Delta}}\right]^{-1}\,. \tag{17}$$

in which the signs "$\pm$" correspond to the cases of ingoing and outgoing particles.

IOP Conf. Series: Journal of Physics: Conf. Series **942** (2017) 012008 doi:10.1088/1742-6596/942/1/012008

**ii)** For the case of massless particles, considering the $s$-wave approximation ($L = 0$) and setting $K = 0$ yields

$$\dot{\bar{r}} = \Delta \left(\sqrt{1-\Delta} \pm 1\right)^{-1} = \pm 1 - \sqrt{1-\Delta}\,. \tag{18}$$

Remarkably, the massless particle's geodesics equation (18) coincides with the result in Ref. [9].

So far, the previous shortcomings in deriving the geodesics have been tamed, and we have achieved the geodesic equations of the charged massive and massless particles tunneling across the Reissner-Nordström AdS black hole in a unified and self-consistent way.

## 4. Tunneling radiation across the event horizon

Hawking describes black hole radiation as a quantum tunneling process triggered by vacuum fluctuations [15]. Based on this scenario, a pair of particles is created near the horizon. The positive energy particle tunnels out to infinity while the negative energy "partner" is absorbed into the black hole and arrives at the singularity point after passing through the one-way membrane. The energy of the black hole decreases and the horizon shrinks. Here we take the self-gravitation between particles into consideration and permit the mass fluctuation of black holes under the conditions of energy conservation. Assuming that a particle with the energy $\omega$ and charge $q$ tunnels out the black hole, the mass and electric charge of the black hole system would change to $(M - \omega)$ and $(Q - q)$, respectively. Meanwhile, the line element and the vector potential switch to

$$d\bar{s}^2 = -\bar{\Delta} dt^2 + 2\sqrt{1-\bar{\Delta}} dt dr + dr^2 + r^2 d\theta^2 + r^2 \sin^2\theta d\phi^2\,, \tag{19}$$

$$F = d\bar{A}\,, \quad \bar{A} = -\frac{Q-q}{r} dt\,, \tag{20}$$

where $\bar{\Delta} = 1 - 2\left(M-\omega\right)/r + \left(Q-q\right)^2/r^2 - \left(\Lambda/3\right) r^2$.

According to the WKB approximation, the emission rate of the tunneling particles is closely related to the action of the tunneling process, that is

$$\Gamma \sim e^{-2\mathrm{Im}S}\,. \tag{21}$$

As alluded before, the generalized coordinate $\mathcal{A}_t$ is an ignorable one. Therefore, to eliminate this degree of freedom, we can express the action as

$$\begin{aligned} S &= \int_{t_i}^{t_f} \left(\mathcal{L} - P_{\mathcal{A}_t} \dot{\bar{\mathcal{A}}}_t\right) dt = \int_{r_i}^{r_f} \left(P_r \dot{\bar{r}} - P_{\mathcal{A}_t} \dot{\bar{\mathcal{A}}}_t\right) \frac{dr}{\dot{\bar{r}}} \\ &= \int_{r_i}^{r_f} \left[\int_{(0,0)}^{(P_r, P_{\mathcal{A}_t})} \left(\dot{\bar{r}}\; dP_r' - \dot{\bar{\mathcal{A}}}_t\; dP_{\mathcal{A}_t}'\right)\right] \frac{dr}{\dot{\bar{r}}}\,, \end{aligned} \tag{22}$$

where, $r_i$ and $r_f$ denote the horizon radii before and after the shrinkage, $P_r$ and $P_{\mathcal{A}_t}$ are the canonical momenta conjugated to $r$ and $\mathcal{A}_t$. Proceeding with an explicit calculation, we substitute the Hamilton's equations of motion

$$\dot{\bar{r}} = \left.\frac{d\mathcal{H}}{dP_r}\right|_{(r;\ \mathcal{A}_t, P_{\mathcal{A}_t})} = \frac{d\left(M-\omega\right)}{dP_r}\,, \tag{23}$$

$$\dot{\bar{\mathcal{A}}}_t = \left.\frac{d\mathcal{H}}{dP_{\mathcal{A}_t}}\right|_{(\mathcal{A}_t;\ r, P_r)} = \Phi \frac{d\left(Q-q\right)}{dP_{\mathcal{A}_t}}\,, \tag{24}$$

IOP Conf. Series: Journal of Physics: Conf. Series **942** (2017) 012008 doi:10.1088/1742-6596/942/1/012008

into Eq. (22) and have

$$S = \int_{r_i}^{r_f} \int_{(M,Q)}^{(M-\omega,Q-q)} \left[ d\left(M-\omega'\right) - \Phi' d\left(Q-q'\right) \right] \frac{dr}{\bar{r}} \,. \tag{25}$$

In general, it is a crucial challenge to finish the tunneling integration in the study of black hole radiation via tunneling. As a highlight, here we tactfully utilize the thermodynamic properties of the black hole to simplify the calculation.

To facilitate completing the above integration, we need the help of the asymptotic behavior of the outgoing particles near the event horizon

$$\bar{r} \approx \kappa' \left(r - r'_+\right), \tag{26}$$

where $\kappa'$ represents the surface gravity on the horizon after the particle emission. Moreover, it can be verified that thermodynamical quantities, such as the entropy and electric potential

$$S' = \pi {r'_+}^2, \quad \Phi' = \frac{Q-q'}{r'_+}, \tag{27}$$

satisfy the differential form of the first law of black hole thermodynamics as below

$$d(M-\omega') = \frac{\kappa'}{2\pi} dS' + \Phi' d(Q-q')\,. \tag{28}$$

With these relations in hand, the imaginary part of the action can be calculated as

$$\begin{aligned} \mathrm{Im}S &\approx \mathrm{Im} \int_{r_i}^{r_f} \int_{(M,Q)}^{(M-\omega,Q-q)} \frac{d\left(M-\omega'\right) - \Phi' d\left(Q-q'\right)}{\kappa' \left(r - r'_+\right)} dr \\ &= -\frac{1}{2} \int_{S_{BH}(M,Q)}^{S_{BH}(M-\omega,Q-q)} dS' = -\frac{1}{2} \Delta S_{BH} \,. \end{aligned} \tag{29}$$

Note that the above integral was worked out by deforming the contour around the single pole $r = r'_+$. We switched the order of integration and first carried out the integral with respect to $r$. Eventually, the emission rate of the charged massive particles tunneling from the Reissner-Nordström black hole in the context of AdS spacetime is

$$\Gamma \sim e^{-2\mathrm{Im}S} = e^{\Delta S_{BH}}, \tag{30}$$

in which $\Delta S_{BH} = S_{BH}(M-\omega, Q-q) - S_{BH}(M,Q)$. As a result, the tunneling probability of a Reissner-Nordström AdS black hole is closely linked with the change of Bekenstein-Hawking entropy and the real radiation spectrum deviates from the pure thermal spectrum.

## 5. Summary

In this work, we have remedied the urgent shortcomings in deriving the geodesics of tunneling particles. By applying the Lagrangian analysis on the action, utilizing the conserved constants and the 4-velocity normalization condition, we have redefined the geodesic equations. Eventually, the geodesic equations of massive and massless particles have been worked out in a unified and self-consistent manner. Furthermore, by employing the prevailing Parikh-Wilczek's semi-classical tunneling method, we have clarified that, even if in the context of AdS spacetime, the tunneling probability of Reissner-Nordström black holes is still closely related to the change of Bekenstein-Hawking entropy and the real radiation spectrum deviates from the pure thermal one. It is noteworthy that, in contrast to the conventional treatment frequently used, the superiority of the calculation resorting to the first law of black hole thermodynamics during the tunneling integration is highlighted.

## Acknowledgements
I would like to thank the organizers of the 3rd Karl Schwarzschild Meeting for an excellent scientific conference and warm hospitality. I am grateful to Prof. Martin Ammon for useful guide during the characteristic "Mentor Program" and Prof. Juan Maldacena for enlightening discussions during the Coffee break.

## References
[1] S.W. Hawking, *Nature* **248**, 30 (1974)
[2] S.W. Hawking, *Commun. Math. Phys.* **43**, 199 (1975)
[3] E.C. Vagenas, *Phys. Lett. B* **559**, 65 (2003). arXiv:hep-th/0209185
[4] S. Sarkar, D. Kothawala, *Phys. Lett. B* **659**, 683 (2008). arXiv:0709.4448
[5] R. Banerjee, B.R. Majhi, E.C. Vagenas, *Phys. Lett. B* **686**, 279 (2010). arXiv:0907.4271
[6] P. Nicolini, E. Winstanley, *JHEP* **1111**, 075 (2011). arXiv:1108.4419
[7] G.M. Deng, *Gen. Rel. Grav.* **46**, 1757 (2014). arXiv:1705.04922
[8] Y.G. Miao, Z.M. Xu, *Phys. Lett. B* **772**, 542 (2017). arXiv:1704.07086
[9] M.K. Parikh, F. Wilczek, *Phys. Rev. Lett.* **85**, 5042 (2000). arXiv:hep-th/9907001
[10] K. Nozari, S.H. Mehdipour, *JHEP* **0903**, 061 (2009). arXiv:0902.1945
[11] K. Matsuno, K. Umetsu, *Phys. Rev. D* **83**, 064016 (2011). arXiv:1101.2091
[12] L.D. Landau, E.M. Lifshitz, *Classical Theory of Fields* (Pergamon Press, London, 1987)
[13] J.Y. Zhang, Z. Zhao, *JHEP* **0510**, 055 (2005)
[14] S. Chandrasekhar, *The Mathematical Theory of Black Holes* (Oxford University Press, New York, 1983)
[15] J.B. Hartle, S.W. Hawking, *Phys. Rev. D* **13**, 2188 (1976)

IOP Conf. Series: Journal of Physics: Conf. Series **942** (2017) 012009 doi:10.1088/1742-6596/942/1/012009

# QED loop effects in the spacetime background of a Schwarzschild black hole

**Viacheslav A. Emelyanov**
Institute of Theoretical Physics, Karlsruhe Institute of Technology, 76131 Karlsruhe, Germany
E-mail: viacheslav.emelyanov@kit.edu

**Abstract.** The black-hole evaporation implies that the quantum-field propagators in a local Minkowski frame acquire a correction, which gives rise to this process. The modification of the propagators causes, in turn, non-trivial local effects due to the radiative/loop diagrams in non-linear QFTs. In particular, there should be imprints of the evaporation in QED, if one goes beyond the tree-level approximation. Of special interest in this respect is the region near the black-hole horizon, which, already at tree level, appears to show highly non-classical features, e.g., negative energy density and energy flux into the black hole.

## 1. Introduction

After many years of theoretical and experimental studies, we have been successful in establishing a couple of models, which incredibly well describe physics at microscopic and macroscopic scales. The basic ideas behind of the Standard Model of Particle Physics and Cosmology, are different, but not inconsistent. It is the Einstein equivalence principle that allows us to go over from the general relativity regime to the particle physics one.

According to the large scale observations, the global geometrical structure of the Universe turns out to be roughly described by de Sitter spacetime. At smaller length scales, the Universe is no longer homogeneous and isotropic as a consequence of the presence of Dark Matter as well as baryonic matter in the form of clusters of galaxies, galaxies and so on. However, Minkowski spacetime appears to be a good approximation to the Universe locally, i.e. in the neighbourhood of any given space-time point.

The geometrical structure relevant for Particle Physics is given by the Minkowski metric of a local Minkowski frame. The Poincaré group $\mathrm{R}^4 \rtimes \mathrm{SO}(1,3)$ is, thus, a local isometry group of the Universe. This group is employed to define the notion of particles, which are associated with its unitary and irreducible representations, with a no-particle state being the local Minkowski vacuum. The Lehmann-Symanzik-Zimmermann reduction formula relates, in turn, physical particle states with poles of the quantum-field propagators. We know *a posteriori* that this formalism adequately describes various reactions observed in the particle colliders on Earth.

The notion of particles identified with the unitary, irreducible representations of the Poincaré group and, hence, labeled by values of its two Casimir generators, i.e. mass and spin, is the only one, which has been successfully tested so far. The Wigner particle is, thus, a localized object, that can be defined in a local Minkowski frame only. It seems that if the local curvature in a certain space-time region of the Universe becomes comparable with, e.g., the Compton length $\lambda_c$ of the electron, then one should not expect that the notion of the electron is physically

IOP Conf. Series: Journal of Physics: Conf. Series **942** (2017) 012009 doi:10.1088/1742-6596/942/1/012009

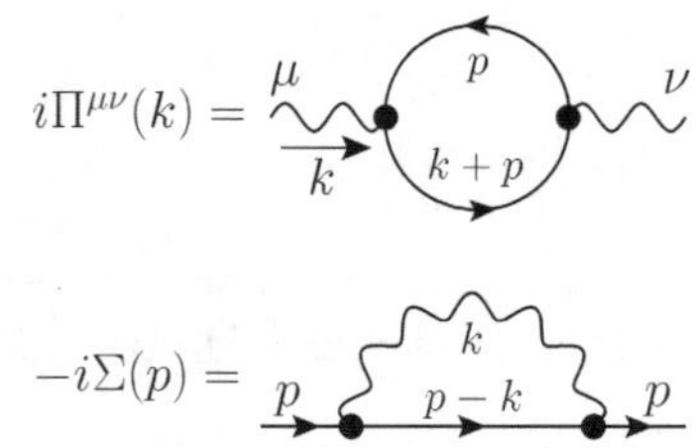


**Figure 1.** One-loop diagrams providing the lowest order correction to the photon as well as electron self-energy. The object $\Pi^{\mu\nu}(k)$ is known as the vacuum polarization tensor. The vacuum polarization gives rise to the running of the electric charge and contributes to the Lamb shift.

meaningful there. Nevertheless, the Wigner electrons (i.e. electrons we observe in the particle colliders) might, still, make physical sense if relativistic, i.e. correspond to the de Broglie length $\lambda_e \ll \lambda_c$. Therefore, the Wigner electrons have to be highly relativistic in that region. Although the proof of this conjecture in a lab is impossible, we take it for granted below as a natural extrapolation of physics we have experimentally probed up to now.

The fact that we find ourselves in non-flat spacetime implies that one should bear in mind the curvature length scale $l_c$. Still, the Minkowski-space approximation employed in Particle Physics is adequate, whenever the length scale of a certain particle reaction is negligible with respect to $l_c$. Consequently, one does not need to integrate over all space-time points of the Universe in the coordinate representation of vertices in non-linear QFTs. Moreover, the same also appears to hold in the path integral formalism and, furthermore, one can use the powerful technique of the Fourier transform as if the Universe were globally Minkowski spacetime.

Thus, one can imagine that the local curvature at a given point of the Universe is described by the length scale $l_c$. According to the Einstein equivalence principle, one can introduce a local Minkowski frame in its vicinity. In this frame, we can follow standard methods used in Particle Physics to quantize, e.g., the Maxwell or Dirac field. There is no reason to expect that this procedure is physically meaningless, because otherwise it would be unnatural to assume that the formalism works on Earth only. Of course, one should pay attention to the fact that the Standard Model is not scale invariant. For instance, if $l_c \lesssim \lambda_c \propto 1/m_e$, one should then allow only electrons/positrons as external legs in the Feynman diagrams with momenta $|\vec{p}| \gg 1/l_c$. In this paper, we intend to employ these ideas near a black hole, so that $l_c$ corresponds to the size of the black-hole event horizon, i.e. $l_c = r_H = 2M$, where $M$ is a black-hole mass.

Throughout this paper we assume that $c = G = k_B = \hbar = 1$, unless otherwise stated.

## 2. Loop effects in Quantum Electrodynamics

A plenty of observable effects in QFT cannot be understood at tree level, i.e. without counting loop diagrams. In quantum electrodynamics, the most remarkable manifestations of the loops are the anomalous magnetic moment of the electron, the running of the electric charge and the Lamb shift. We shall deal in this paper with the one-loop correction to the photon and electron self-energy (see Fig. 1).

### *2.1. Isotropic electron-positron plasma*

In the presence of particles and/or external fields, novel non-trivial manifestations of the loops appear. As an instructive example, we consider first an isotropic and neutral plasma composed of electrons, positrons and photons.

IOP Conf. Series: Journal of Physics: Conf. Series **942** (2017) 012009 doi:10.1088/1742-6596/942/1/012009

*Maxwell- and Dirac-field propagators:* To determine how the plasma particles influence, e.g, the photon kinematics, one needs to know the vector- and fermion-field propagators inside the plasma. One can show that

$$\underbrace{\frac{i(\not{p}+m)}{p^2-m^2+i\varepsilon}}_{\text{Vacuum}} \quad \Longrightarrow \quad \underbrace{\frac{i(\not{p}+m)}{p^2-m^2+i\varepsilon} - 2\pi\,\frac{(\not{p}+m)}{e^{\beta|p_0|}+1}\,\delta(p^2-m^2)}_{\text{Isotropic plasma}}\,, \tag{1}$$

where $\beta \equiv 1/T$ is the inverse plasma temperature, and

$$\underbrace{\frac{-i\eta_{\mu\nu}}{k^2+i\varepsilon}}_{\text{Vacuum}} \quad \Longrightarrow \quad \underbrace{\frac{-i\eta_{\mu\nu}}{k^2+i\varepsilon} - 2\pi\,\frac{\eta_{\mu\nu}}{e^{\beta|k_0|}-1}\,\delta(k^2)}_{\text{Isotropic plasma}}\,. \tag{2}$$

Since the electron-positron field is massive, the extra term in its propagator is not suppressed by the Boltzmann factor $\exp(-\beta m_e)$ if we consider the case $T \gg m_e$. It is legitimate, then, to neglect the mass parameter $m_e$, so that the fermion field becomes effectively massless. This limit is known as the hard-thermal-loop approximation [13].

*Modified dispersion relation:* The one-loop effects originating from non-vanishing temperature are the Debye screening of a point-like charge, emergent propagating degrees of freedom (so-called plasmon and plasmino) as a result of the collective plasma-particles excitations and others (see, e.g., [13]). We focus herein on the photon and electron dispersion relations.

With the propagators given in Eqs. (1) and (2), we find in the on-mass-shell limit in the hot plasma ($T \gg m_e$) that

$$k_0^2 - \vec{k}^2 \;\approx\; \frac{2\pi}{3}\,\alpha T^2\,, \tag{3}$$

$$p_0^2 - \vec{p}^2 \;\approx\; m_e^2 + \pi\alpha T^2 \tag{4}$$

at one-loop level, where $\alpha \equiv e^2/4\pi$ is the fine-structure constant and $m_e$ is a physical electron mass (whereas $m$ is a bare electron mass). Thus, QED photons acquire a thermal gauge-invariant mass, which changes their kinematics [13].

### *2.2. Anisotropic electron-positron plasma*

An anisotropic plasma is characterized by a unit vector $\vec{n}$, that gives the direction along which the plasma moves. This medium appears to be closely related to the quantum physics near to and far from the black-hole horizon [4, 5, 7, 8].

*Maxwell- and Dirac-field propagators:* If the anisotropic and neutral electron-positron plasma is held at temperature $T$, then one needs to implement in Eqs. (1) and (2) the following changes:

$$\delta(k^2) \quad \Longrightarrow \quad 2\pi|\vec{k}|\,\delta(\vec{k}-k_0\vec{n})\,, \tag{5}$$

$$\delta(p^2-m^2) \quad \Longrightarrow \quad 2\pi|\vec{p}|\,\delta(\vec{p}-(p_0^2-m^2)^{\frac{1}{2}}\vec{n})\,, \tag{6}$$

in order to gain the photon- and electron-positron-field propagators in this medium.

Starting with quantum kinetic theory (see, e.g., [6]), it is straightforward to show that the anisotropic plasma is described by the following one-particle distribution:

$$f(x,p) \;=\; \frac{g_s}{2\pi^2}\,\frac{|\vec{p}|^2}{e^{\beta p_0}-(-1)^{2s}}\,\theta(\vec{p}\cdot\vec{n})\delta(\vec{p}\times\vec{n})\,, \tag{7}$$

IOP Conf. Series: Journal of Physics: Conf. Series **942** (2017) 012009 doi:10.1088/1742-6596/942/1/012009

where $p_0$ is either $|\vec{p}|$ or $(|\vec{p}|^2+m_s^2)^{\frac{1}{2}}$, depending on whether the quantum field of spin $s$ is massless or massive and $g_s$ is its number of degrees of freedom. For instance, this implies that the photon number current $\vec{j}_\gamma$ is given by $n_\gamma{\cdot}\vec{n}$, where $n_\gamma = 2\zeta(3)T^3/\pi^2$ is the photon number density in the plasma. Therefore, the anisotropy of the plasma manifests itself in the macroscopic local observables.

*Modified dispersion relation:* It turns out that the modified dispersion relations given in Eqs. (3) and (4) are independent on whether the electron-positron plasma is isotropic or anisotropic. This holds only at one-loop approximation. The photon dispersion relation at two-loop order gets a dependence on the unit vector $\vec{n}$, implying that the two-loop diagrams do not contribute to the thermal photon mass [9].

### *2.3. Near-horizon region of small black holes*

The near-horizon region of black holes is exotic from a classical point of view. Specifically, the renormalized stress tensor of the quantum fields describes an inward flux of the negative energy density, originating well outside of the black-hole horizon [16, 2, 12]. This explains, however, why the black-hole horizon shrinks, i.e. $2\dot{M} < 0$, where dot refers to the Schwarzschild time coordinate.

This non-classical flux, like the Hawking flux in the far-from-horizon region, is anisotropic. This phenomenon is described by the field propagators of the form given in Eqs. (1) and (2) with the modifications (5) and (6) implemented with an extra multiple factor roughly equaling to $-27/4$ and $\beta = 2/T_H$ [6, 7],[1] where $T_H$ is the Hawking temperature. Thus, the following result holds:

$$k_0^2 - \vec{k}^2 \approx -\frac{9\pi}{8}\,\alpha T_H^2\,, \tag{8}$$

$$p_0^2 - \vec{p}^2 \approx m_e^2 - \frac{27\pi}{16}\,\alpha T_H^2\,, \tag{9}$$

for black holes with $T_H \gg m_e$ [8].

## 3. Concluding remarks

We have found above that $k^2 < 0$ and $p^2 < 0$ close to the event horizon. What is the physics of this result?

*Einstein acausality vs. instability:* First, it is useful to consider a classical field model with a "wrong" sign in the mass-squared term. In general, there are two logically non-excludable ways of quantizing this theory. One can either discard modes with imaginary frequencies or take those into account. In the former case, the Einstein causality is violated, whereas, in the latter one, it is not, but the price is the quantum-vacuum instability [3, 15, 1]. The very reason of the instability are frequency modes with $|\vec{q}| < \mu$, where $\mu > 0$ is the field "mass". In other words, the modes with $q_0^2 = \vec{q}^2 - \mu^2 < 0$ are exponentially increasing in time. This model, however, is not equivalent to our case, because $m_\gamma^2 < 0$ is a one-loop result obtained within standard QED, but in the non-standard quantum-vacuum state.

It is further instructive to consider a classical field model with the standard mass term, but, e.g., in flat de Sitter spacetime with the Hubble constant $H$. In this case, the superhorizon modes increase without bound at future time infinity. These correspond to imaginary time-dependent frequencies. This fact does not serve any problem in Particle Physics, because those modes have

[1] The factor 2 in $\beta$ comes from the Fermi normal coordinates, which are introduced in the near-horizon region.

a physical wavelength larger than the space-time curvature, i.e. no notion of particles can be associated with them [14].

It, now, seems that the tachyonic scalar field model with $\mu \lesssim H$ could be physically (almost) indistinguishable from a massless scalar field in local tests in the particle colliders. Still, the dispersion relation is $q_0^2 = \vec{q}^2 - \mu^2$ with the IR cutoff $l_{IR} \ll 1/H \lesssim 1/\mu$ and, hence, $|\vec{q}| \gg H \gtrsim \mu$. This, in turn, means that $dq_0/d|\vec{q}| \approx 1 + \mu^2/2\vec{q}^2 > 1$ holds for scalar particles. Moreover, any local perturbation (e.g., a particle) is oblivious to the large-scale structure and, thus, described by momenta with $|\vec{q}|$ much larger than $1/l_{IR}$,[2] implying that this perturbation is described by the group velocity larger than $c$.

This last example reflects our case, if one takes into account that $m_\gamma^2 < 0$ emerges due to the black-hole evaporation at one-loop approximation. Hence, it does not hold at length scales larger than $r_H$, because the field propagators, we have employed in the one-loop computations, are accurate in the local Minkowski frame only. Thus, a high-energy QED photon/electron might be able to propagate out of the black hole, bringing information about its internal structure. This appears to provide a semi-classical mechanism for the proposal of [10, 11] towards the resolution of the information-loss problem.

*Black-hole explosions:* The non-violent Einstein acausality near the event horizon implies that the black-hole explosions happen much earlier, i.e. rather before black holes reach the Planckian size. It might provide the explanation of why these explosions have not been observed so far, because these appear to be less energetic events in the sky than it has been thought.

## Acknowledgments

It is a pleasure to thank the participants of the 3rd Karl Schwarzschild Meeting on Gravitational Physics and the Gauge/Gravity Correspondence for interesting discussions and the organizers for making it all happen. I am also thankful to Frans Klinkhamer and José Queiruga for their comments on an early version of this paper.

## References

[1] Aharonov, Y., Komar. A., Susskind L.: Superluminal behavior, causality, and instability. Phys. Rev. **182**, 1400 (1969)
[2] Candelas, P.: Vacuum polarization in Schwarzschild space-time. Phys. Rev. D **21**, 2185 (1980)
[3] Dhar, J., Sudarshan, E.C.G.: Quantum field theory of interacting tachyons. Phys. Rev. **174**, 1808 (1968)
[4] Emelyanov, V.A.: Effective photon mass from black-hole formation. Nucl. Phys. B **919**, 110 (2017)
[5] Emelyanov, V.A.: Quantum electromagnetic phenomena far from small evaporating black holes. In: Gomboc A. (ed.) New Frontiers in Black Hole Astrophysics, pp. 279-282. Cambridge University Press (2016)
[6] Emelyanov, V.A.: Quantum kinetic theory of a massless scalar model in the presence of a Schwarzschild black hole. Annalen Phys. **529**, 1700078 (2017)
[7] Emelyanov, V.A.: Magnetostatic-field screening induced by small black holes. Nucl. Phys. B **921**, 796 (2017)
[8] Emelyanov, V.A.: Can a QED photon propagate out of the black-hole horizon? arXiv:hep-th/1703.05078
[9] Emelyanov, V.A.: Two-loop QED photon self-energy near the black-hole horizon. In progress
[10] Giddings, S.B.: Models for unitary black hole disintegration. Phys. Rev. D **85**, 044038 (2012)
[11] Giddings, S.B.: Nonviolent nonlocality. Phys. Rev. D **88**, 064023 (2013)
[12] Giddings, S.B.: Hawking radiation, the Stefan-Boltzmann law, and unitarization. Phys. Lett. B **754**, 39 (2016)
[13] Le Bellac, M.: Thermal Field Theory. Cambridge University Press (1996)
[14] Mukhanov, V.F., Winitzki, S.: Introduction to Quantum Effects in Gravity. Cambridge University Press (2007)

[2] It should be noted that the (standard) Fourier transform is inapplicable at length scales, at which the space-time curvature is non-negligible. This is, however, not a problem in Particle Physics, because particles are localized in coordinate space. The infinite-extent approximation of a local Minkowski frame is, thus, applicable, if particles are characterized by a high-enough energy scale.

IOP Conf. Series: Journal of Physics: Conf. Series **942** (2017) 012009 doi:10.1088/1742-6596/942/1/012009

[15] Schroer, B.: Quantization of $m^2 < 0$ field equations. Phys. Rev. D **3**, 1764 (1971)
[16] Unruh, W.G.: Origin of the particles in black hole evaporation. Phys. Rev. D **15**, 365 (1977)

IOP Conf. Series: Journal of Physics: Conf. Series **942** (2017) 012010 doi:10.1088/1742-6596/942/1/012010

# Time dependence of entanglement for steady state formation in $AdS_3/CFT_2$

**M Flory**[1], **J Erdmenger**[2], **D Fernández**[3], **E Megías**[4], **A Straub**[5] **and P Witkowski**[6]

[1] Institute of Physics, Jagiellonian University, Łojasiewicza 11, 30-348 Kraków, Poland
[2] Institut für Theoretische Physik und Astrophysik, Julius-Maximilians-Universität Würzburg, Am Hubland, 97074 Würzburg, Germany
[3] University of Iceland, Science Institute, Dunhaga 3, 107 Reykjavík, Iceland
[4] Departamento de Física Teórica, Universidad del País Vasco UPV/EHU, Apartado 644, 48080 Bilbao, Spain
[5] Max-Planck-Institut für Physik (Werner-Heisenberg-Institut), Föhringer Ring 6, 80805 München, Germany
[6] Max-Planck-Institut für Physik komplexer Systeme, Nöthnitzer Straße 38, D-01187 Dresden, Germany

E-mail: mflory@th.if.uj.edu.pl

**Abstract.** We consider a holographic model of two 1+1-dimensional heat baths at different temperatures joined at time $t = 0$, such that a steady state heat-current region forms and expands in space for times $t > 0$. After commenting on the causal structure of the dual 2+1-dimensional spacetime, we present how to calculate the time-dependent entanglement entropy of the boundary system holographically. We observe that the increase rate of the entanglement entropy satisfies certain bounds known from the literature on entanglement tsunamis. Furthermore, we check the validity of several non-trivial entanglement inequalities in this dynamic system.

## 1. Introduction and setup

During the recent years, there has been a number of fruitful attempts to apply methods of gauge/gravity duality to the study of strongly coupled far-from-equilibrium systems. In these proceedings, following our recent paper [1], we will investigate the specific holographic system studied in [2]. There, a setup was proposed where at time $t = 0$, two semi-infinite heat reservoirs are brought into contact, leading to a temperature profile of the form

$$T(t = 0, x) = T_{\rm L}\,\theta(-x) + T_{\rm R}\,\theta(x), \tag{1}$$

where $x$ is a spacial coordinate. In what follows, we will focus on the case where the field theory is $1 + 1$ dimensional, leading to a $2 + 1$ dimensional holographic bulk description.

On the field theory side, it can be shown that as time evolves onward from the initial condition (1), a constant energy flow $\langle J_E \rangle \neq 0$, the *steady state*, develops in a growing region between two shockwaves emerging at the speed of light from $t = x = 0$ [2], see figure 1. In contrast to similar setups in higher dimensions, this $1 + 1$ dimensional setup has the special property that the shock waves with which the steady state region expands are dissipation-free, meaning that

IOP Conf. Series: Journal of Physics: Conf. Series **942** (2017) 012010 doi:10.1088/1742-6596/942/1/012010

the temperature profile evolving from the initial conditions (1) will involve sharp step-functions even for times $t \gg 0$.

**Figure 1.** At $t = 0$, both systems are isolated and independently at equilibrium. Evolving forward in time from $t = 0$, a spatially homogeneous non-equilibrium steady state develops in the middle region, carrying an energy current $J_E$.

Our goal will be to use holographic methods [3, 4, 5] in order to calculate the time dependence of entanglement entropy in this system. The holographic dual to the $1+1$ dimensional large $c$ CFT described above is given by an asymptotically $AdS_3$ (vacuum) solution to Einsteins equations which generically takes the form [2, 6]

$$ds^2 = \frac{1}{z^2}\left(dz^2 + g_{\mu\nu}(z,x,t)dx^\mu dx^\nu\right) , \tag{2a}$$

$$g_{tt}(z,x,t) = -\left[1 - z^2\left(f_R(x-t) + f_L(x+t)\right)\right]^2 + \left[z^2\left(f_R(x-t) - f_L(x+t)\right)\right]^2 , \tag{2b}$$

$$g_{tx}(z,x,t) = -2z^2\left(f_R(x-t) - f_L(x+t)\right) , \tag{2c}$$

$$g_{xx}(z,x,t) = \left[1 + z^2\left(f_R(x-t) + f_L(x+t)\right)\right]^2 - \left[z^2\left(f_R(x-t) - f_L(x+t)\right)\right]^2 , \tag{2d}$$

where we have set the AdS-radius $L \equiv 1$ and $f_R(v)$ and $f_L(v)$ are arbitrary functions to be determined from the initial conditions. Specifically, demanding (1) to hold in the dual CFT enforces the choice [2]

$$f_{L/R}(v) \equiv \frac{\pi^2}{2}\left(T_L^2 + \left(T_R^2 - T_L^2\right)\theta(v)\right) . \tag{3}$$

The appearance of the step function in (3) means that the bulk spacetime will appear to consist of several sectors describing static and boosted BTZ black holes, matched together along co-dimension one hypersurfaces. This was discussed in more details in [1], see also figure 2. The important thing to notice is that these matching hypersurfaces are spacelike in nature, hence they cannot be interpreted as physical objects carrying information in the bulk.

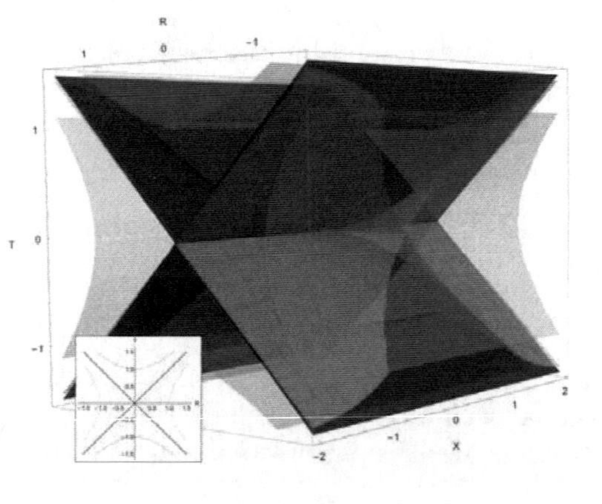


**Figure 2.** Three dimensional Kruskal diagram (with Kruskal coordinates $T, R, X$, see [1]) of the bulk spacetime (2a). The black diagonal sheets correspond to the event horizons. The red surfaces show the singularities, and the orange surfaces show the location of the boundary at $z = 0$. The green surface is the worldvolume of the bulk hypersurfaces along which the different regions of spacetime are glued together. A steady state region forms both for $t > 0$ and for $t < 0$. Only the $t > 0$ region is considered to be physically relevant.

IOP Conf. Series: Journal of Physics: Conf. Series **942** (2017) 012010 doi:10.1088/1742-6596/942/1/012010

## 2. Entanglement entropy

For static bulk spacetimes, entanglement entropy $S(A) \equiv -\text{Tr}\,[\rho_A \log(\rho_A)]$ for a boundary subregion $A$ with reduced density matrix $\rho_A$ can be calculated via the Ryu-Takayanagi (RT) proposal [3, 4] $S_A(A) = \frac{\mathcal{A}(\gamma_A)}{4G_N}$, where $\mathcal{A}(\gamma_A)$ is the area of a minimal spacelike codimension two bulk surface $\gamma_A$ anchored on the asymptotic boundary. The generalisation of this prescription to time-dependent bulk spacetimes was discussed in [5]. In this Hubeny-Rangamany-Takayanagi (HRT) proposal, the bulk surface $\gamma_A$ is merely required to be an *extremal* hypersurface. In the $2+1$ dimensional bulk setup (2a) of interest here, the curves $\gamma_A$ will be spacelike geodesics.

As discussed in [1], there are two possibilities to tackle the geodesic problem in the background (2a). The seemingly most direct way would be to solve the geodesic equations numerically. Due to the way in which the geodesics have to be anchored at the boundary, this would require the use of non-trivial numerical techniques such as *relaxation methods* (similar to [7]) or *shooting methods* as successfully implemented for the system of our interest in [1, 8, 9].

Due to the piecewise structure of (2a) discussed in section 1, there is however a more elegant way to solve this problem in the system of interest: The spacetime can be divided into four sectors $(t+x>0 \wedge x-t>0)$, $(t+x>0 \wedge x-t<0)$, $(t+x<0 \wedge x-t>0)$ and $(t+x<0 \wedge x-t<0)$, see figure 2. As in each of these quarters the bulk metric is the one of a (boosted) BTZ black hole [10], and as geodesic curves and distances between spacelike separated points in a BTZ spacetime are known analytically [11], it is possible to write the (renormalised) geodesic distance between two given boundary points of the spacetime (2a) as a function of the coordinates of the matching points where the corresponding geodesic passes through the hypersurfaces along which the different sectors are glued onto each other. Extremising the resulting length with respect to the coordinates of the matching points yields then the correct overall geodesic length. This can be seen as a way to derive the appropriate refraction conditions for geodesics at the hypersurfaces. For situations where the geodesic in question crosses the hypersurface only once, this procedure has been worked out in detail in [1]. It was found that the relevant equations can be solved analytically in the cases where either $T_R$ and $T_L$ are perturbatively small or where $T_R = 0$ exactly with finite $T_L$. Otherwise, the numerical problem boils down to finding the solutions to a set of *algebraic* equations.

## 3. Results

### 3.1. General behaviour

Figures 3 and 4 show representative results obtained in [1] with the method described in section 2. There are three important observations to be made: Firstly, the time evolution of entanglement entropy is *causal*, i.e. before the shockwave moving over the boundary with the speed of light from $t = x = 0$ enters the interval and after it has left the interval, the entanglement entropy of a given boundary interval is constant. Only while the shockwave travels through the interval does non-trivial time evolution take place.

Secondly, when at least one of the temperatures $T_L$ and/or $T_R$ is large compared to the interval length $\ell$, the entanglement entropy shows a mostly linear increase as a function of time. This is exemplified in figure 4, where the time dependent entanglement entropy is rescaled as

$$f_A(\rho) \equiv \frac{S_A(t) - S_A(t=0)}{S_A(t=\infty) - S_A(t=0)} \qquad \text{with} \qquad \rho \equiv (t - x_{min})/\Delta t\,, \tag{4}$$

where $\Delta t = \ell = x_{max} - x_{min}$. This rescaling serves to map the time-dependent part of the curve $S_A(t)$ to the unit square of the axes of figure 4. Interestingly, such a linear increase is reminiscent to the similar linear increase of entanglement entropy observed in *entanglement tsunamis* [12, 13, 14, 15, 16, 17, 18, 19]. However, thirdly, when both $T_L$ and $T_R$ are small compared to the interval length $\ell$, we obtain the universal low-temperature approximation

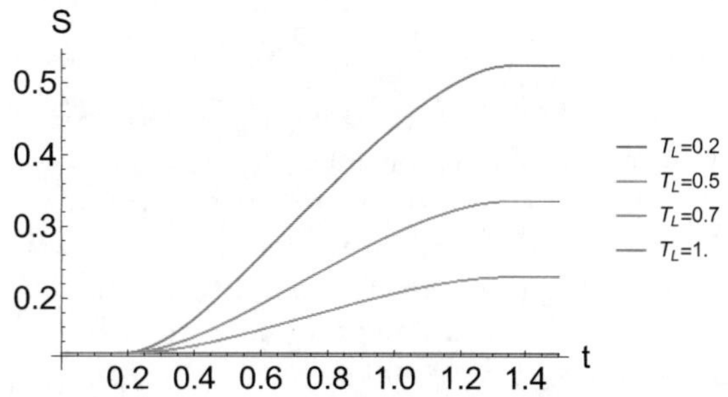


**Figure 3.** Renormalised entanglement entropies $S_A(t)$ for a boundary interval $0.175 \leq x \leq 1.35$, $T_R = 0.195$ and $T_L$ ranging from 0.2 (lowest curve) to 1.0 (highest curve) as a function of boundary time $t$.

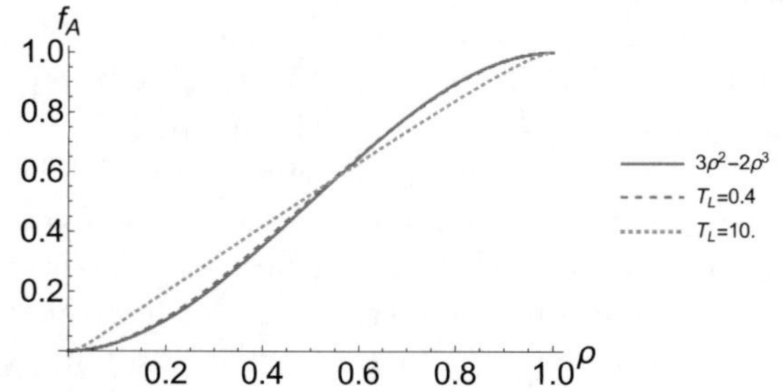


**Figure 4.** Rescaled entanglement entropy $f_A(\rho)$ for the interval of figure 3 with $T_L = 0.4$ and $T_L = 10$ compared to the universal formula (5).

formula [1, 8, 9]

$$f_A(\rho) \simeq 3\rho^2 - 2\rho^3 , \qquad 0 \leq \rho \leq 1 . \tag{5}$$

*3.2. Bounds on entropy increase*

The above mentioned similarity between the time dependence of entanglement entropy in entanglement tsunamis and the system investigated in [1, 2] motivates the study of bounds on the entropy increase rate. To this end, we define the averaged entropy increase rate [1],

$$v_{av} \equiv \frac{\Delta S}{\Delta t} = \frac{L}{4G\ell} \log\left(\frac{T_L \sinh(\pi \ell T_R)}{T_R \sinh(\pi \ell T_L)}\right), \tag{6}$$

and the normalised averaged entropy increase rate

$$\tilde{v}_{av} \equiv \frac{v_{av}}{s_{eq}}, \quad |\tilde{v}_{av}| \leq \left|\frac{T_R - T_L}{T_R + T_L}\right| \leq 1, \tag{7}$$

where $s_{eq} = \frac{L}{4G}\pi(T_L + T_R)$ is the entropy density of the final steady state. The bound shown in (7) is exactly the one expected from the study of entanglent tsunamis in $1 + 1$ dimensional CFTs [12, 13, 14, 15, 16, 17, 18, 19]. Furthermore, we can define the normalised *momentary* entropy increase rate

$$\tilde{v} \equiv \frac{1}{s_{eq}} \frac{dS(\ell, t)}{dt}, \tag{8}$$

which in all cases that we studied analytically or numerically also satisfies the bound $|\tilde{v}| \leq 1$.

*3.3. Entanglement inequalities*

Another area of interest is the validity of *entanglement inequalities.* It is known that the RT prescription for the calculation of entanglement entropy automatically ensures certain inequalities, amongst others the *strong subadditivity* (SSA) [20, 21]

$$S(AB) + S(BC) - S(ABC) - S(B) \geq 0, \tag{9}$$

where e.g. $S(BC)$ is the entanglement entropy of the union of the two boundary regions $A$ and $B$. In the time-dependent case where the HRT prescription applies, these inequalities are also satisfied under the assumption that certain energy conditions hold [22, 23]. Using the RT prescription, several other entanglement inequalities involving $n = 5$ or more intervals have been derived in [24], one of them taking the form

$$\begin{aligned} &S(ABC) + S(BCD) + S(CDE) + S(DEA) + S(EAB) - S(ABCDE) \\ &- S(BC) - S(CD) - S(DE) - S(EA) - S(AB) \geq 0. \end{aligned} \tag{10}$$

The study of such entanglement entropy inequalities in AdS/CFT is of considerable interest for the understanding of holography. For example, the study of such inequalities for time-dependent setups may provide insights into the role of energy conditions in holography [25].

In [1], we hence numerically tested, in the time dependent model of section 1, the validity of (10) and several other entanglement entropy inequalities proven or conjectured to hold for static cases . In order to do so, we had to take into account the different geodesic configurations or *phases* that may play a role in the RT or HRT prescription for the union of $n \geq 2$ boundary regions. See e.g. figure 5 for the case $n = 3$. Our results can be summarised as follows: For $n = 3$ intervals $A$, $B$ and $C$, strong subadditivity (9) is satisfied, as expected based on [22, 23]. In contrast to the situations investigated in [26], even the *engulfed phase* (see figure 5) can be the relevant one in specific examples. Generically, happens when the middle interval is very small compared to the gap between the other two intervals.

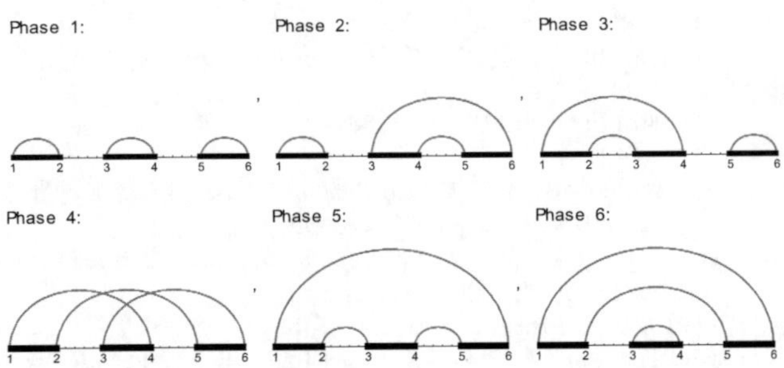


**Figure 5.** 6 possible geodesic configurations for the calculation of entanglement entropy via the HRT proposal for $n = 3$ boundary intervals. The 4th configuration is considered to be unphysical. According to the nomenclature of [26], the phase 6 is referred to as *engulfed* phase.

For $n = 4$ intervals $A$, $B$, $C$ and $D$, the only inequality that we are checking is the *positivity of four-partite information*

$$S(A) + S(B) + ... - S(AB) - S(AC) - ... + S(ABC) + S(ABC) + ... - S(ABCD) \geq 0, \tag{11}$$

which was conjectured to hold in [27, 28]. We, however, find a number of examples for sets of four intervals where this inequality is violated. As it was pointed out in [29] and was explicitly checked by us, this happens already in holographic systems with static bulk-spacetime duals.

For $n = 5$ intervals $A$, $B$, $C$, $D$ and $E$, we find numerous violations of the *negativity of five-partite information* [27, 28], see the similar discussion for $n = 4$. Furthermore, we check the inequality (10) as well as the other inequalities for $n = 5$ intervals derived in [24]. The result is that we find *not a single case* in which any of these inequalities is violated, see for example figure 6. We view this as a clear indication that these inequalities (such as (10)), although so far only proven in the static case, will generally also hold in physical time-dependent cases.

**Acknowledgments**
MF was supported by NCN grant 2012/06/A/ST2/00396. We would also like to thank the organisers of the Karl Schwarzschild meeting 2017. EM is supported by the U. País Vasco UPV/EHU, Spain, as a Visiting Professor.

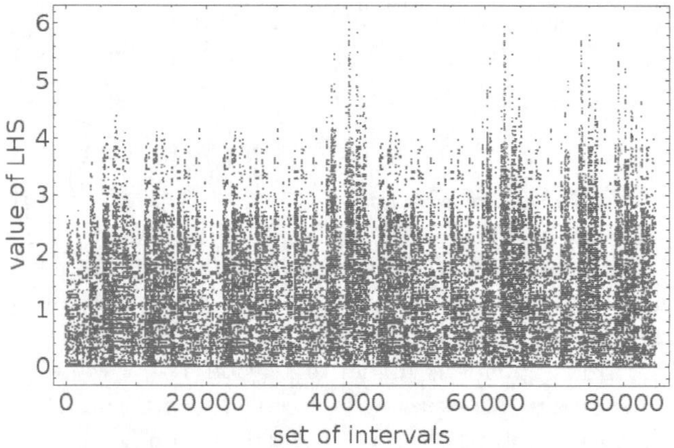


**Figure 6.** The left-hand sides of the inequality (10) for 84579 different sets of five intervals for which this quantity does not vanish trivially. Clearly there are no violations of the inequality (10). The different colors in the figure stand for different permutations of the inequality, obtained from (10) by exchanging the labels ($A$, $B$, ...) of the intervals. We are here displaying results for the example where $T_L = 9$, $T_R = 1$ and boundary time $t = 1$.

## References

[1] Erdmenger J, Fernandez D, Flory M, Megias E, Straub A K and Witkowski P 2017 (*Preprint* 1705.04696)
[2] Bhaseen M J, Doyon B, Lucas A and Schalm K 2013 [Nature Phys.11,5(2015)] (*Preprint* 1311.3655)
[3] Ryu S and Takayanagi T 2006 *Phys. Rev. Lett.* **96** 181602 (*Preprint* hep-th/0603001)
[4] Ryu S and Takayanagi T 2006 *JHEP* **08** 045 (*Preprint* hep-th/0605073)
[5] Hubeny V E, Rangamani M and Takayanagi T 2007 *JHEP* **07** 062 (*Preprint* 0705.0016)
[6] Banados M 1998 Three-dimensional quantum geometry and black holes *Trends in theoretical physics II. Proceedings, 2nd La Plata Meeting, Buenos Aires, Argentina, November 29-December 4, 1998* pp 147–169 [AIP Conf. Proc.484,147(1999)] (*Preprint* hep-th/9901148)
[7] Ecker C, Grumiller D and Stricker S A 2015 *JHEP* **07** 146 (*Preprint* 1506.02658)
[8] Megias E 2015 *PoS* **EPS-HEP2015** 366 (*Preprint* 1510.04219)
[9] Megias E Far-from-equilibrium energy flow and entanglement entropy *5th International Conference on New Frontiers in Physics Kolymbar, Crete, Greece, July 6-14, 2016 (Preprint* 1701.00098*)*
[10] Banados M, Teitelboim C and Zanelli J 1992 *Phys. Rev. Lett.* **69** 1849–1851 (*Preprint* hep-th/9204099)
[11] Shenker S H and Stanford D 2014 *JHEP* **03** 067 (*Preprint* 1306.0622)
[12] Abajo-Arrastia J, Aparicio J and Lopez E 2010 *JHEP* **11** 149 (*Preprint* 1006.4090)
[13] Balasubramanian V, Bernamonti A, de Boer J, Copland N, Craps B, Keski-Vakkuri E, Muller B, Schafer A, Shigemori M and Staessens W 2011 *Phys. Rev. Lett.* **106** 191601 (*Preprint* 1012.4753)
[14] Balasubramanian V, Bernamonti A, de Boer J, Copland N, Craps B, Keski-Vakkuri E, Muller B, Schafer A, Shigemori M and Staessens W 2011 *Phys. Rev.* **D84** 026010 (*Preprint* 1103.2683)
[15] Balasubramanian V, Bernamonti A, Copland N, Craps B and Galli F 2011 *Phys. Rev.* **D84** 105017 (*Preprint* 1110.0488)
[16] Hartman T and Maldacena J 2013 *JHEP* **05** 014 (*Preprint* 1303.1080)
[17] Liu H and Suh S J 2014 *Phys. Rev. Lett.* **112** 011601 (*Preprint* 1305.7244)
[18] Li Y Z, Wu S F, Wang Y Q and Yang G H 2013 *JHEP* **09** 057 (*Preprint* 1306.0210)
[19] Liu H and Suh S J 2014 *Phys. Rev.* **D89** 066012 (*Preprint* 1311.1200)
[20] Lieb E H and Ruskai M B 1973 *Journal of Mathematical Physics* **14** 1938–1941 (*Preprint* http://dx.doi.org/10.1063/1.1666274)
[21] Headrick M and Takayanagi T 2007 *Phys. Rev.* **D76** 106013 (*Preprint* 0704.3719)
[22] Wall A C 2014 *Class. Quant. Grav.* **31** 225007 (*Preprint* 1211.3494)
[23] Prudenziati A 2016 *Phys. Rev.* **D93** 106003 (*Preprint* 1509.07885)
[24] Bao N, Nezami S, Ooguri H, Stoica B, Sully J and Walter M 2015 *JHEP* **09** 130 (*Preprint* 1505.07839)
[25] Lashkari N, Rabideau C, Sabella-Garnier P and Van Raamsdonk M 2015 *JHEP* **06** 067 (*Preprint* 1412.3514)
[26] Ben-Ami O, Carmi D and Sonnenschein J 2014 *JHEP* **11** 144 (*Preprint* 1409.6305)
[27] Alishahiha M, Mozaffar M R M and Tanhayi M R 2015 *JHEP* **09** 165 (*Preprint* 1406.7677)
[28] Mirabi S, Tanhayi M R and Vazirian R 2016 *Phys. Rev.* **D93** 104049 (*Preprint* 1603.00184)
[29] Hayden P, Headrick M and Maloney A 2013 *Phys. Rev.* **D87** 046003 (*Preprint* 1107.2940)

IOP Conf. Series: Journal of Physics: Conf. Series **942** (2017) 012011 doi:10.1088/1742-6596/942/1/012011

# Gravity's Rainbow and Black Hole Entropy

**Remo Garattini**

**Abstract.** We consider the effects of Gravity's Rainbow on the computation of black hole entropy using a dynamical brick wall model. An explicit dependence of the radial coordinate approaching the horizon is proposed to analyze the behavior of the divergence. We find that, due to the modification of the density of states, the brick wall can be eliminated. The calculation is extended to include rotations and in particular to a Kerr black hole in a comoving frame.

Università degli Studi di Bergamo, Facoltà di Ingegneria,
Viale Marconi 5, 24044 Dalmine (Bergamo) ITALY.
INFN - sezione di Milano, Via Celoria 16, Milan, Italy.

E-mail: remo.garattini@unibg.it

## 1. Introduction

In recent years many attempts to modify gravity at the fundamental level have been proposed. Some of them, like $f(R)$ theories[1], have been considered to basically change the large scale structure of the space-time and some others have been conceived to modify the short scale behavior. Among them, Gravity's Rainbow (GRw) seems to be promising in dealing with Ultra-Violet divergences (UV). Indeed, in a series of papers GRw has been used to avoid any regularization/renormalization scheme which appear in conventional Quantum Field Theory calculations like one loop corrections to classical quantities[2]. This amazing property has been applied also to black holes and in particular to the computation of black hole entropy[3]. In this last case, the idea is to avoid to introduce a cut-off of Planckian size known as "*brick wall*"[4]. The "*brick wall*" appears when one uses a statistical mechanical approach to explain the famous Bekenstein-Hawking formula[5, 6]

$$S_{BH} = \frac{1}{4}A/l_P^2, \tag{1}$$

relating the entropy of a black hole and its area. Indeed, when one tries to adopt such an approach, one realizes that the density of energy levels of single-particle excitations is divergent near the horizon. Of course, several attempts have been done to avoid the introduction of the *brick wall*. For instance, without modifying gravity at any scale, it has been suggested that the *brick wall* could be absorbed in a renormalization of Newton's constant[7, 8, 9], while other authors approached the problem of the divergent brick wall using Pauli-Villars regularization[10, 11, 12]. Other than GRw other proposals have been made in the context of modified gravity. For instance, non-commutative geometry introduces a natural thickness of the horizon replacing the 't Hooft's brick wall[16] and *Generalized Uncertainty Principle* (GUP) modifies the Liouville measure[13, 14, 15]. To understand how GRw works we need to define two unknown functions $g_1(E/E_P)$ and $g_2(E/E_P)$ having the following property

$$\lim_{E/E_P\to 0} g_1(E/E_P) = 1 \qquad \text{and} \qquad \lim_{E/E_P\to 0} g_2(E/E_P) = 1. \tag{2}$$

In this formalism introduced by Magueijo and Smolin[18], the Einstein's field equations are replaced by a one parameter family of equations

$$G_{\mu\nu}(E) = 8\pi G(E) T_{\mu\nu}(E) + g_{\mu\nu}\Lambda(E), \tag{3}$$

where $G(E)$ is an energy dependent Newton's constant and $\Lambda(E)$ is an energy dependent cosmological constant, respectively. They are defined so that $G(0)$ is the physical Newton's constant and $\Lambda(0)$ is the usual cosmological constant. In this context, the Schwarzschild solution of (3) becomes

$$ds^2(E) = -\left(1 - \frac{2MG(0)}{r}\right)\frac{dt^2}{g_1^2(E/E_P)} + \frac{dr^2}{\left(1 - \frac{2MG(0)}{r}\right) g_2^2(E/E_P)} + \frac{r^2}{g_2^2(E/E_P)}\left(d\theta^2 + \sin^2\theta d\phi^2\right) \tag{4}$$

and it can be easily generalized in the following way

$$ds^2(E) = -\exp(-2A(r))\left(1 - \frac{b(r)}{r}\right)\frac{dt^2}{g_1^2(E/E_P)} + \frac{dr^2}{\left(1 - \frac{b(r)}{r}\right) g_2^2(E/E_P)} + \frac{r^2}{g_2^2(E/E_P)} d\Omega^2. \tag{5}$$

The function $b(r)$ will be referred to as the "shape function" and it may be thought of as specifying the shape of the spatial slices. The location of the horizon is determined by the equation $b(r_H) = r_H$. On the other hand, $A(r)$ will be referred to as the "redshift function" and describes how far the total gravitational redshift deviates from that implied by the shape function. The line element (5) describes any spherically symmetric space-time. It is interesting to wonder what happens when one introduces rotations. Rotating black holes have a good description in terms of the Kerr metric which, in the context of GRw, becomes[19]

$$ds^2(E) = \frac{g_{tt}dt^2}{g_1^2(E/E_P)} + \frac{2g_{t\phi}dtd\phi}{g_1(E/E_P)\,g_2(E/E_P)} + \frac{g_{\phi\phi}d\phi^2}{g_2^2(E/E_P)} + \frac{g_{rr}dr^2}{g_2^2(E/E_P)} + \frac{g_{\theta\theta}d\theta^2}{g_2^2(E/E_P)}, \tag{6}$$

where

$$g_{tt} = -\frac{\Delta - a^2\sin^2\theta}{\Sigma}, \qquad g_{t\phi} = -\frac{a\sin^2\theta\left(r^2 + a^2 - \Delta\right)}{\Sigma},$$
$$g_{\phi\phi} = \frac{\left(r^2 + a^2\right)^2 - \Delta a^2\sin^2\theta}{\Sigma}\sin^2\theta, \qquad g_{rr} = \frac{\Sigma}{\Delta}, \qquad g_{\theta\theta} = \Sigma, \tag{7}$$

and

$$\Delta = r^2 - 2MGr + a^2, \qquad \Sigma = r^2 + a^2\cos^2\theta. \tag{8}$$

Here $M$ and $a$ are mass and angular momentum per unit mass of the black hole, respectively. $\Delta$ vanishes when $r = r_\pm = MG \pm \sqrt{(MG)^2 - a^2}$, while $g_{tt}$ vanishes when $r = r_{S\pm} = MG \pm \sqrt{(MG)^2 - a^2\cos^2\theta}$: they are not modified by GRw and the outer horizon or simply horizon is located at $r_+ = r_H$. Note that the Kerr metric modified by GRw (6) reduces to the standard rotating black hole background when $E/E_P \to 0$. In this contribution we will consider the effect of GRw on Black Hole entropy computation even for a rotating background. Units in which $\hbar = c = k = 1$ are used throughout the paper.

IOP Conf. Series: Journal of Physics: Conf. Series **942** (2017) 012011 doi:10.1088/1742-6596/942/1/012011

## 2. GRw Entropy for a Schwarzschild Black Hole

To see what happens in practice for the Schwarzschild Black Hole, we define a real massless scalar field whose Euler-Lagrange equations are

$$\frac{1}{\sqrt{-g}}\partial_\mu\left(\sqrt{-g}g^{\mu\nu}\partial_\nu\right)\phi=0. \tag{9}$$

The formalism has been outlined in detail in [3] and therefore we refer the reader to [3] for details. If $\phi$ has the separable form

$$\phi\left(t,r,\theta,\varphi\right)=\exp\left(-iEt\right)Y_{lm}(\theta,\varphi)f\left(r\right), \tag{10}$$

then the equation for $f\left(r\right)$ reads

$$\left[\frac{g_2^2\left(E/E_P\right)\exp\left(A\left(r\right)\right)}{r^2}\partial_r\left(r^2\exp\left(-A\left(r\right)\right)\left(1-\frac{b\left(r\right)}{r}\right)\partial_r\right)-V_l\left(r\right)\right]f_{nl}=0, \tag{11}$$

where

$$V_l\left(r\right)=\left(\frac{l(l+1)}{r^2}-\frac{E_{nl}^2g_1^2\left(E/E_P\right)\exp\left(2A\left(r\right)\right)}{1-\frac{b(r)}{r}}\right) \tag{12}$$

and where $Y_{lm}(\theta,\varphi)$ is the usual spherical harmonic function. In order to use the WKB approximation to compute the entropy, we define the following r-dependent radial wave number $k(r,l,E)$

$$k_r^2(r,l,E)\equiv\frac{1}{\left(1-\frac{b(r)}{r}\right)}\left[\exp\left(2A\left(r\right)\right)\frac{E^2h^2\left(E/E_P\right)}{\left(1-\frac{b(r)}{r}\right)}-\frac{l(l+1)}{r^2}\right], \tag{13}$$

where

$$h\left(E/E_P\right)=\frac{g_1\left(E/E_P\right)}{g_2\left(E/E_P\right)}. \tag{14}$$

The number of modes with frequency less than $E$ is given approximately by

$$n(E)=\frac{1}{\pi}\int_0^{l_{max}}(2l+1)\int_{r_H}^{R}\sqrt{k^2(r,l,E)}drdl, \tag{15}$$

Here it is understood that the integration with respect to $r$ and $l$ is taken over those values which satisfy $r_H\leq r\leq R$ and $k^2(r,l,E)\geq 0$. Thus, from Eq.(13) we get

$$\frac{dn(E)}{dE}=\frac{2}{\pi}\frac{d}{dE}\left(\frac{1}{3}E^3h^3\left(E/E_P\right)\right)\int_{r_H}^{R}dr\frac{\exp\left(3A\left(r\right)\right)}{\left(1-\frac{b(r)}{r}\right)^2}r^2. \tag{16}$$

Since the free energy can be written as

$$F=\frac{1}{\beta}\int_0^\infty\ln\left(1-e^{-\beta E}\right)\frac{dn(E)}{dE}dE, \tag{17}$$

where $\beta$ is the inverse temperature measured at infinity. Plugging Eq.(16) into (17) we find

$$F_{r_H}=\frac{2}{\pi}\frac{1}{\beta}\int_0^\infty\ln\left(1-e^{-\beta E}\right)\frac{d}{dE}\left(\frac{1}{3}E^3h^3\left(E/E_P\right)\right)\left[\int_{r_H}^{r_1}drr^2\frac{\exp\left(3A\left(r\right)\right)}{\left(1-\frac{b(r)}{r}\right)^2}\right]dE \tag{18}$$

and

$$F_R = \frac{2}{\pi}\frac{1}{\beta}\int_0^\infty \ln\left(1-e^{-\beta E}\right)\frac{d}{dE}\left(\frac{1}{3}E^3h^3\left(E/E_P\right)\right)\left[\int_{r_1}^{R} drr^2\frac{\exp\left(3A\left(r\right)\right)}{\left(1-\frac{b(r)}{r}\right)^2}\right]dE. \tag{19}$$

Assuming that $A\left(r\right)<\infty$, $\forall r\in\left[r_H,+\infty\right)$, $F_R$ is dominated by large volume effects for large $R$ and it will give the contribution to the entropy of a homogeneous quantum gas in flat space at a uniform temperature $T$ when GRw is considered. If GRw does not come into play, then the radial part of $F_{r_H}$ becomes divergent in proximity of $r_H$. On the other hand, if we allow that the "*brick wall* $r_0$" be affected by GRw, namely $r_0\equiv r_0\left(E/E_P\right)$, then the radial integration $F_{r_H}$ becomes

$$\int_{r_H+r_0(E/E_P)}^{r_1} drr^2\frac{\exp\left(3\Lambda\left(r\right)\right)}{\left(1-\frac{b(r)}{r}\right)^2}\simeq r_H^4\frac{\exp\left(3A\left(r_H\right)\right)}{\left(1-b'\left(r_H\right)\right)^2}\frac{1}{r_0\left(E/E_P\right)}$$

$$= r_H^3\frac{\exp\left(3A\left(r_H\right)\right)}{\left(1-b'\left(r_H\right)\right)^2}\frac{1}{\sigma\left(E/E_P\right)}, \tag{20}$$

where we have assumed that, in proximity of the throat the brick wall can be written as $r_0\left(E/E_P\right)=r_H\sigma\left(E/E_P\right)$ with

$$\sigma\left(E/E_P\right)\to 0,\qquad E/E_P\to 0. \tag{21}$$

Plugging Eq.(20) into $F_{r_H}$ we obtain, after an integration by parts

$$F_{r_H} = -\frac{C_{r_H}}{3\beta r_H}\int_0^\infty \frac{E^3h^3\left(E/E_P\right)}{\sigma\left(E/E_P\right)}\left[\frac{\beta}{\left(\exp\left(\beta E\right)-1\right)}-\frac{\ln\left(1-e^{-\beta E}\right)}{E_P\sigma\left(E/E_P\right)}\sigma'\left(E/E_P\right)\right]. \tag{22}$$

$h\left(E/E_P\right)$ must be chosen in such a way to allow the convergence when $E/E_P\to\infty$, thus we assume that

$$h\left(E/E_P\right)=\exp\left(-\frac{E}{E_P}\right)\qquad\text{and}\qquad \sigma\left(E/E_P\right)=h^\delta\left(E/E_P\right)\left(\frac{E}{E_P}\right)^\alpha. \tag{23}$$

In particular, for $\delta=0$; $\alpha=2$, one finds that the entropy becomes

$$S=\beta^2\frac{\partial F_{r_H}}{\partial\beta}\underset{\beta E_P\gg 1}{\simeq}\frac{A_{r_H}E_P^2}{4}\frac{\exp\left(2A\left(r_H\right)\right)}{1-b'\left(r_H\right)}\frac{2}{9\pi}. \tag{24}$$

where we have used the expression for the surface gravity in the low energy limit. As we can see the "*brick wall*" does not appear.

## 3. GRw Entropy for the Kerr Black Hole

To discuss the entropy for a Kerr black hole we have two options: we can use a rest observer at infinity (ROI) or we can use a Zero Angular Momentum Observer (ZAMO)[20, 21]. The ROI frame is described by the line element (6) and the appropriate form of the free energy is the following

$$F=\frac{1}{\beta}\int_0^\infty dn\left(E\right)\ln\left(1-e^{-\beta(E-m\Omega)}\right). \tag{25}$$

IOP Conf. Series: Journal of Physics: Conf. Series **942** (2017) 012011 doi:10.1088/1742-6596/942/1/012011

It is immediate to see that when we use a ROI, the problem of superradiance appears when the free energy (25) is computed in the range $0 < E < m\Omega$. On the other hand when a ZAMO is considered, the free energy (25) becomes similar to the one used for a Schwarzschild black hole (17). Basically this happens because near the horizon the metric becomes

$$ds^2 = -\frac{N^2dt^2}{g_1^2\left(E/E_P\right)} + g_{\phi\phi}\frac{d\phi^2}{g_2^2\left(E/E_P\right)} + g_{rr}\frac{dr^2}{g_2^2\left(E/E_P\right)} + g_{\theta\theta}\frac{d\theta^2}{g_2^2\left(E/E_P\right)} \tag{26}$$

and the mixing between $t$ and $\phi$ disappears. Moreover when we use a ZAMO frame, the superradiance does not come into play because there is no ergoregion. Indeed since we have defined

$$N^2 = g_{tt} - \frac{g_{t\phi}^2}{g_{\phi\phi}} = -\frac{1}{g^{tt}} = -\frac{\Delta\sin^2\theta}{g_{\phi\phi}}, \tag{27}$$

$N^2$ vanishes when $r \to r_H$. Therefore if we repeat the same steps which led us to the computation of (16), one finds

$$\frac{dn(E)}{dE} = \frac{1}{8\pi^2}\int d\theta d\bar{\phi}\int_{r_H}^{R} dr\left(-g^{tt}\right)^{\frac{3}{2}}\sqrt{g_{rr}g_{\theta\theta}g_{\phi\phi}}\frac{1}{3}\frac{d}{dE}\left(h^3\left(E/E_P\right)E^3\right), \tag{28}$$

where the solution of the massless Klein-Gordon equation (9) assumes the form

$$\phi\left(x\right) = \exp\left(-iEt + im + iK\left(r,\theta\right)\right) \tag{29}$$

with

$$k_r = \frac{\partial K\left(r,\theta\right)}{\partial r}, \qquad k_\theta = \frac{\partial K\left(r,\theta\right)}{\partial\theta} \tag{30}$$

defined in such a way to use the WKB approximation. In proximity of the horizon, the free energy can be approximated by

$$F_{r_H} = \frac{1}{8\pi^2\beta}\int d\theta d\bar{\phi}\int_0^\infty \ln\left(1-e^{-\beta E}\right)\frac{d}{dE}\left(\frac{1}{3}h^3\left(E/E_P\right)E^3\right)dE\int_{r_H}^{r_1} dr\left(-g^{tt}\right)^{\frac{3}{2}}\sqrt{g_{rr}g_{\theta\theta}g_{\phi\phi}} \tag{31}$$

which can be further reduced to

$$F_{r_H} \simeq \frac{C\left(r_H,\theta\right)}{8\pi^2\beta}\int_0^\infty \frac{\ln\left(1-e^{-\beta E}\right)}{\sigma\left(E/E_P\right)}\frac{d}{dE}\left(\frac{1}{3}h^3\left(E/E_P\right)E^3\right)dE, \tag{32}$$

where

$$C\left(r_H,\theta\right) = \int d\theta d\bar{\phi}\left[\frac{\left(r_H^2+a^2\right)^4\sin\theta}{r_H\left(r_H-r_-\right)^2\Sigma_H}\right]. \tag{33}$$

With an integration by parts one finds

$$F_{r_H} = -\frac{C\left(r_H,\theta\right)}{24\pi^2\beta}\int_0^\infty \frac{E^3h^3\left(E/E_P\right)}{\sigma\left(E/E_P\right)}\left[\frac{\beta}{\left(\exp\left(\beta E\right)-1\right)} - \frac{\ln\left(1-e^{-\beta E}\right)}{E_P\sigma\left(E/E_P\right)}\sigma'\left(E/E_P\right)\right]dE. \tag{34}$$

If we adopt the same choice of the previous section described by (23) and we fix our attention on the particular values $\delta = 0$ and $\alpha = 2$, one finds

$$\begin{aligned} F_{r_H} &= -\frac{C\left(r_H,\theta\right)}{24\pi^2\beta}\int_0^\infty\left[\frac{\beta E e^{-3E/E_P}}{\left(\exp\left(\beta E\right)-1\right)} - 2e^{-3E/E_P}\ln\left(1-e^{-\beta E}\right)\right]dE \\ &= -\frac{C\left(r_H,\theta\right)}{24\pi^2\beta}\left[\zeta\left(2,1+\frac{3}{\beta E_P}\right) + \frac{\beta E_P}{3}\left(\gamma+\Psi\left(1+\frac{3}{\beta E_P}\right)\right)\right], \end{aligned} \tag{35}$$

IOP Conf. Series: Journal of Physics: Conf. Series **942** (2017) 012011 doi:10.1088/1742-6596/942/1/012011

where $\zeta(s,\nu)$ is the Hurwitz zeta function, $\Gamma(x)$ is the gamma function and $\Psi(x)$ is the digamma function. In the limit where $\beta E_P \gg 1$, at the leading order, one finds that the entropy can be approximated by

$$S = \beta^2 \frac{\partial F_{rw}}{\partial \beta} = \frac{E_P^2}{36\beta} \int d\theta d\bar{\phi} \left[ \frac{\left(r_H^2 + a^2\right)^4 \sin\theta}{r_H \left(r_H - r_-\right)^2 \Sigma_H} \right] \tag{36}$$

and even in this case the "*brick wall*" does not appear. Of course the entropy (36) can always be cast in the familiar form

$$S = \frac{A_H}{4G}, \tag{37}$$

where $A_H$ is the horizon area. To summarize, we have shown that the ability of Gravity's Rainbow to keep under control the UV divergences applies also to rotations. However the connection between a ROI and a ZAMO has to be investigated with care[22]. Indeed in the ROI frame, the superradiance phenomenon appears, while in the ZAMO frame does not. Once the connection is established nothing forbids to extend this result to other rotating configuration like, for example, Kerr-Newman or Kerr-Newman-De Sitter (Anti-De Sitter).

**Acknowledgments**
The author would like to thank MDPI for a partial financial support.

## References

[1] S. Capozziello and M. De Laurentis, *Phys. Rept.* **509** (2011); arXiv:1108.6266 [gr-qc].
[2] R. Garattini and G. Mandanici, *Phys. Rev.* **D 83**, 084021 (2011); arXiv:1102.3803 [gr-qc]. R. Garattini and G. Mandanici, *Phys. Rev.* **D 85**, 023507 (2012); arXiv:1109.6563 [gr-qc]. R. Garattini, *JCAP* **017** 1306, (2013); arXiv:1210.7760 [gr-qc]; R. Garattini and B. Majumder, *Nucl. Phys.* **B 884** 125, (2014); arXiv:1311.1747 [gr-qc]. R. Garattini and B. Majumder, *Nucl. Phys.* **B 883** 598, (2014); arXiv:1305.3390 [gr-qc]. R. Garattini and E. N. Saridakis, *Eur. Phys. J.* **C 75**, 343 (2015); ArXiv:1411.7257 [gr-qc].
[3] R. Garattini, *Phys. Lett.* **B 685**, 329 (2010); arXiv:0902.3927 [gr-qc].
[4] G. 't Hooft, *Nucl. Phys.* **B 256**, 727 (1985).
[5] J. D. Bekenstein, *Phys. Rev.* **D 7**, 949 (1973).
[6] S. W. Hawking, *Comm. Math. Phys.* **43**, 199 (1975).
[7] L. Susskind and J. Uglum, *Phys. Rev.* **D 50**, 2700 (1994).
[8] J.L.F. Barbon and R. Emparan, *Phys. Rev.* **D 52**, 4527 (1995), hep-th/9502155.
[9] E. Winstanley, *Phys. Rev.* **D 63**, 084013 (2001), hep-th/0011176.
[10] J.- G. Demers, R. Lafrance and R.C. Myers, *Phys. Rev.* **D 52**, 2245 (1995), gr-qc/9503003.
[11] D. V. Fursaev and S. N. Solodukhin, *Phys. Lett.* **B 365**, 51 (1996), hep-th/9412020.
[12] S. P. Kim, S. K. Kim, K.-S. Soh and Jae Hyung Yee, *Int. J. Mod. Phys.* **A 12**, 5223 (1997), gr-qc/9607019.
[13] X. Li, *Phys. Lett.* **B 540**, 9 (2002), gr-qc/0204029.
[14] Z. Ren, W. Yue-Qin and Z. Li-Chun, Class. Quant. Grav. **20** (2003), 4885.
[15] G. Amelino-Camelia, *Class.Quant.Grav.* **23**, 2585 (2006), gr-qc/0506110. G. Amelino-Camelia, *Gen.Rel.Grav.* **33**, 2101 (2001), gr-qc/0106080.
[16] H. Bai and M. L. Yan, JHEP **7** 58 (2003), gr-qc/0303006.
[17] J. L. Cortes and J. Gamboa, *Phys. Rev.* **D 71**, 065015 (2005); hep-th/0405285
[18] J. Magueijo and L. Smolin, *Class. Quant. Grav.* **21**, 1725 (2004) [arXiv:gr-qc/0305055].
[19] Y. Zhao and X. Liu, arXiv:1606.06285 [gr-qc].
[20] E. Chang-Young, D. Lee and M. Yoon, *Class. Quant. Grav.* **26**, 155011 (2009) [arXiv:0811.3294 [hep-th]].
[21] M. H. Lee and J. K. Kim, *Phys. Rev.* **D 54**, 3904 (1996); ArXiv:hep-th/9603055.
[22] In preparation.

IOP Conf. Series: Journal of Physics: Conf. Series **942** (2017) 012012 doi:10.1088/1742-6596/942/1/012012

# Quantum post-Newtonian theory for corpuscular Black Holes

**Andrea Giugno**

Arnold Sommerfeld Center, Ludwig-Maximilians-Universität, Theresienstraße 37, 80333 München, Germany

E-mail: `A.Giugno@physik.uni-muenchen.de`

**Abstract.** We discuss an effective theory for the quantum static gravitational potential in spherical symmetry up to the first post-Newtonian correction. We build a suitable Lagrangian from the weak field limit of the Einstein-Hilbert action coupled to pressureless matter. Classical solutions of the field equation lead to the correct post-Newtonian expansion. Furthermore, we portray the Newtonian results in a quantum framework by means of a coherent quantum state, which is properly corrected to accomodate post-Newtonian corrections. These considerations provide a link between the corpuscular model of Dvali and Gomez and standard post-Newtonian gravity, laying the foundations for future research.

## 1. Introduction

Newtonian theory describes gravity in terms of forces, allowing to well-define energy and to provide the existence of a related scalar potential. On the other hand, to specify these quantities in General Relativity (GR) proves to be a harder task, since it characterizes the gravitational field through the local geometry of space-time [1]. Therefore, the notion of a scalar potential is not univocal for any observer. In the post-Newtonian approximation, the local curvature is weak and the velocities of any test particle are non-relativistic, which supports the derivation of an effective theory for the gravitational potential generated by static and isotropic compact sources. This formulation sees the approximated geodesic equation as a standard Newton's law, whose potential is in turn determined by the Poisson equation. We further enrich the picture by adding non-linearities in the quantum interpretation of the gravitational potential [2], in the light of the results of Ref. [3]. It is important to stress out that this "Newtonian-like" approach is one of the fundamental ingredients, which allow to quantise gravitational degrees of freedom with standard methods [4, 5, 6, 7], once some properties like spherical symmetry are implied. The conceptual arena where this study is carried on, is the corpuscular model of gravity brought to light by Dvali and Gomez [8, 9, 10, 11, 12, 13]. A black hole is naturally formed by a large number of gravitons, which superimpose in the same quantum state and hence realise a Bose-Einstein Condensate (BEC) stably on the verge of a quantum phase transition [14, 15, 16, 17, 18, 19, 20]. In addition to that, the gravitons are expected to be marginally bound in their (gravitational) confining potential well [21, 22, 23], whose size is set by the distinctive Compton-de Broglie wavelength $\lambda_{\rm G} \sim R_{\rm H}$, where

$$R_{\rm H} = 2\, l_{\rm p}\, \frac{M}{m_{\rm p}} \tag{1}$$

IOP Conf. Series: Journal of Physics: Conf. Series **942** (2017) 012012 doi:10.1088/1742-6596/942/1/012012

is the gravitational radius of the black hole of ADM mass $M$, and whose depth is directly proportional to the number $N_{\rm G} \sim M^2/m_{\rm p}^2$ of soft quanta [24, 25, 26, 27, 28]. We mention here that we shall work in units of the Planck length $l_{\rm p}$ and mass $m_{\rm p}$, so that the Newton constant reads $G_{\rm N} = l_{\rm p}/m_{\rm p}$, while $\hbar = l_{\rm p}\, m_{\rm p}$. The speed of light is instead normalised to $c = 1$.

Although the original outline [11, 12] only took the degrees of freedom of gravity into consideration, especially when $R_{\rm H}$ is of astrophysical size [29, 30, 31], and substantially neglected the contributions of collapsing baryonic matter, the post-Newtonian approximation [3] emerges in a simple fashion when they are properly taken into account. The concept is straightforward: let us pretend to have $N$ baryons of rest mass $\mu$ very far apart, with their total ADM mass [32] thus given by $M = N\,\mu$. During the gravitational collapse, the baryons are enclosed in a sphere of radius $R$ and possess a (negative) gravitational energy $U_{\rm BG} \sim N\,\mu\, V_{\rm N} \sim -\frac{l_{\rm p}\, M^2}{m_{\rm p}\, R}$, where $V_{\rm N} \sim -\frac{l_{\rm p}\, M}{m_{\rm p}\, R}$ is the associated Newtonian potential. On a quantum mechanical perspective, the link with the classical potential $V_{\rm N}$ is achieved thanks to the expectation value of a scalar field $\hat{\Phi}$, $V_{\rm N} \sim \langle g|\,\hat{\Phi}\,|g\rangle$, over a coherent state $|g\rangle$. This feature entails that the graviton number $N_{\rm G}$ is determined by the normalisation of the coherent state and follows Bekenstein's area law [33], $N_{\rm G} \sim R_{\rm H}^2/l_{\rm p}^2$, when $R = R_{\rm H}$. Moreover, the Compton-de Broglie length $\lambda_{\rm G} \sim R$ is recovered through the assumption that it is the wave-length of almost the entire amount of gravitons. Thus, the (negative) energy of any constituent yields consistently $\epsilon_{\rm G} \sim \frac{U_{\rm BG}}{N_{\rm G}} \sim -\frac{m_{\rm p}\, l_{\rm p}}{R}$ and the graviton self-interaction energy, $U_{\rm GG}(R) \sim N_{\rm G}\,\epsilon_{\rm G}\,\langle g|\,\hat{\Phi}\,|g\rangle \sim \frac{l_{\rm p}^2\, M^3}{m_{\rm p}^2\, R^2}$ reproduces the typical post-Newtonian correction. It is easy to match this machinery with the standard knowledge. Considering indeed a star of size $R \gg R_{\rm H}$ we get $U_{\rm GG} \ll |U_{\rm BG}|$, whereas we recover the so-called "maximal packing" condition of Ref. [8],

$$U(R_{\rm H}) \equiv U_{\rm BG}(R_{\rm H}) + U_{\rm GG}(R_{\rm H}) \simeq 0 \ , \tag{2}$$

when gravity is strongly coupled, *i.e.* $R \simeq R_{\rm H}$.

In the following, we first derive a consistent effective theory for a static and spherically symmetric potential by considering the Einstein-Hilbert action in the aforementioned non-relativistic and weak field regimes. The inclusion of Next-to-Linear Order (NLO) terms provides classical results in compliance with the usual post-Newtonian expansion of the Schwarzschild metric. Furthermore, the quantum state of the soft scalar gravitons is correctly represented by a coherent state, which establishes a link between a microscopic description of gravity and the macroscopic geometry of space-time. Therefore, such an outcome [2] enriches and improves the conclusions of Ref. [3].

## 2. Classical effective theory

In order to build a correct effective theory, we first have to show how a real scalar field can describe the post-Newtonian approximation of the weak field limit of Schwarzschild metric [1]. The reader shall bear in mind, however, that this construction assumes that a specific reference frame has been chosen. The starting point of our discussion is the Einstein-Hilbert action

$$S = \int {\rm d}^4x\,\sqrt{-g}\left(-\frac{m_{\rm p}}{16\,\pi\, l_{\rm p}}\,\mathcal{R} + \mathcal{L}_{\rm M}\right) \ , \tag{3}$$

coupled to the Lagrangian density $\mathcal{L}_{\rm M}$ that represents the ordinary matter, which collapses and acts as a source for the graviational field. We labeled the Ricci scalar with $\mathcal{R}$. As we already stressed out, in order to retrieve the well-known post-Newtonian result the local curvature has to be small, which means that the metric can be expanded as $g_{\mu\nu} = \eta_{\mu\nu} + h_{\mu\nu}$, where $|h_{\mu\nu}| \ll 1$, and $\eta_{\mu\nu} = {\rm diag}(-1, +1, +1, +1)$. Moreover, the characteristic velocity of the matter under

IOP Conf. Series: Journal of Physics: Conf. Series **942** (2017) 012012 doi:10.1088/1742-6596/942/1/012012

consideration is many orders of magnitude smaller than the speed of light in the considered reference frame $x^\mu = (t, \mathbf{x})$. We can safely describe the entire gravitational system through one relevant component of $g_{\mu\nu}$, $h_{00}(\mathbf{x})$, which is also time-independent. The stress-energy tensor is hence completely specified by the non-relativistic energy density, $T_{\mu\nu} \simeq u_\mu\, u_\nu\, \rho(\mathbf{x})$, where $u^\mu = \delta^\mu_0$ represents the four-velocity of the constituents of the source. Furthermore, this tensor allows to choose the matter Lagrangian simply as $\mathcal{L}_{\rm M} \simeq -\rho(\mathbf{x})$, which is enough to our purpose, since we suppose that matter pressure is negligible [34, 35] and we neglect the associated dynamics. We can finally make the identification $h_{00} = -2\, V_{\rm N}$, since the the Newtonian potential $V_{\rm N}$ is known to satisfy the Poisson equation

$$\triangle V_{\rm N} = 4\,\pi\,\frac{l_{\rm p}}{m_{\rm p}}\,\rho\ , \tag{4}$$

while the de Donder gauge fixing takes the very simple form $\triangle h_{00}(\mathbf{x}) = -8\,\pi\,\frac{l_{\rm p}}{m_{\rm p}}\,\rho(\mathbf{x})$. It is now straightforward to introduce an effective scalar field theory for the gravitational potential. To do so, we first insert $V_{\rm N}$ in the Einsten-Hilbert action, and then we compute the related Hamiltonian by making a Legendre transformation of the Lagrangian $L[V_{\rm N}]$. Then, we include non-linearities through the introduction of a self-gravitational source, defined as a the gravitational potential energy $U_{\rm N}$ per unit volume. Leaving a more detailed analysis of such a construction to Ref. [2] (and the Appendices thereof, in particular), here we only report the resulting Lagrangian

$$L[V] = -4\,\pi\int_0^\infty r^2\,{\rm d}r\left[\frac{m_{\rm p}}{8\,\pi\,l_{\rm p}}\left(1-4\,q_\Phi\,V\right)\left(V'\right)^2 + q_{\rm B}\,V\,\rho\left(1-2\,q_\Phi\,V\right)\right]\ . \tag{5}$$

The couplings $q_{\rm B}$ and $q_\Phi$ measure the strength of the interaction of $V$ with baryonic matter and the self-sourcing, respectively. They are rescaled in such a way to recover the known post-Newtonian expansion for $q_{\rm B} = q_\Phi = 1$. The Euler-Lagrange equation for $V$ is given by

$$\triangle V = 4\,\pi\,q_{\rm B}\,\frac{l_{\rm p}}{m_{\rm p}}\,\rho + 2\,q_\Phi\left(1-4\,q_\Phi\,V\right)^{-1}\left(V'\right)^2\ , \tag{6}$$

and it is obviously hard to solve analytically for a general source. We will therefore expand the field $V$ up to first order in the coupling $q_\Phi$, $V(r) = V_{(0)}(r) + q_\Phi\,V_{(1)}(r)$, and solve Eq. (6) order by order. In particular, for the leading order $V_{(0)}$ we have the analog of the Poisson Eq. (4), while

$$\triangle V_{(1)} = 2\left(V'_{(0)}\right)^2 \tag{7}$$

gives the correction at $\mathcal{O}(q_\Phi)$. To linear order in $q_\Phi$, the on-shell Hamiltonian reads

$$H[V] = 2\,\pi\int_0^\infty {\rm d}r\,r^2\left[q_{\rm B}\,\rho\,V\left(1-4\,q_\Phi\,V\right) - q_\Phi\,\frac{3\,m_{\rm p}}{2\,\pi\,l_{\rm p}}\,V\left(V'^2\right)\right] + \mathcal{O}(q_\Phi^2)\ . \tag{8}$$

If we take *e.g.* a matter distribution uniformly distributed within a sphere of radius $R$ [2], not only the solution of the field Equation correctly reproduces the potential that one expects in semiclassical gravity, but also the potential energy computed thanks to Eq. (8) reads

$$U(R) = -q_{\rm B}^2\,\frac{3\,l_{\rm p}\,M^2}{5\,m_{\rm p}\,R} + q_{\rm B}^3\,q_\Phi\,\frac{249\,l_{\rm p}^2\,M^3}{175\,m_{\rm p}^2\,R^2} + \mathcal{O}(q_\Phi^2)\ , \tag{9}$$

$M$ being again the ADM mass sourcing the graviational field up to order $q_\Phi$. It is of paramount importance to notice that $U(R)$ vanishes when $R \simeq 1.2 R_{\rm H}$, realising therefore the "maximal packing" condition (2).

IOP Conf. Series: Journal of Physics: Conf. Series **942** (2017) 012012 doi:10.1088/1742-6596/942/1/012012

## 3. Quantum realisation

In order to canonically quantise the theory, we consider the rescaled real scalar field $\Phi = \sqrt{\frac{m_{\rm p}}{l_{\rm p}}}\, V$, coupled to the static source $J_{\rm B} = 4\,\pi\,\sqrt{\frac{l_{\rm p}}{m_{\rm p}}}\,\rho$ and replace these new quantities in Eq. (5). We straightforwardly obtain the scalar field Lagrangian

$$L[\Phi] = 4\,\pi \int_0^\infty r^2\,{\rm d}r \left[\frac{1}{2}\,\Phi\,\Box\Phi - q_{\rm B}\,J_{\rm B}\,\Phi\left(1 - 2\,q_\Phi\,\sqrt{\frac{l_{\rm p}}{m_{\rm p}}}\,\Phi\right) + 2\,q_\Phi\,\sqrt{\frac{l_{\rm p}}{m_{\rm p}}}\,(\partial_\mu\Phi)^2\,\Phi\right]\;, \tag{10}$$

where we again assumed $\Phi = \Phi(t, r)$. Let us look for a wave-function $|g\rangle$ which reproduces the classical solution of the EOM of $\Phi$. First, we examine the linear case, $q_\Phi = 0$. In terms of the new variables $\Phi$ and $J_{\rm B}$, it is analogous to a Poisson Equation $\triangle\Phi_c(r) = q_{\rm B}\,J_{\rm B}(r)$ , when the field is static as its own source. Upon expanding the EOM on a base of spherically symmetric normal modes, one can solve this Equation in momenta space, so that

$$\tilde{\Phi}_c(k) = -q_{\rm B}\,\frac{\tilde{J}_{\rm B}(k)}{k^2}\;. \tag{11}$$

A shift transformation of the ladder operators related to $\hat{\Phi}$ shows that the coherent state $\hat{a}_k\,|g\rangle = e^{i\,\gamma_k(t)}\,g_k\,|g\rangle$ correctly reproduces the classical solution, $\langle g|\,\hat{\Phi}(t, r)\,|g\rangle = \Phi_c(r)$, with

$$g_k = -q_{\rm B}\,\frac{\tilde{J}_{\rm B}(k)}{\sqrt{2\,l_{\rm p} m_{\rm p}\,k^3}}\;, \tag{12}$$

and $\gamma_k(t) = -kt$. It is possible to read off the mean number of quanta $N_{\rm G}$ from the normalisation condition $\langle g|g\rangle = 1$, which turns out to be

$$N_{\rm G} = \langle g|\int_0^\infty \frac{k^2\,{\rm d}k}{2\,\pi^2}\,\hat{a}_k^\dagger\,\hat{a}_k\,|g\rangle = \int_0^\infty \frac{k^2\,{\rm d}k}{2\,\pi^2}\,g_k^2\;, \tag{13}$$

and $N_{\rm G}$ is shown to precisely equal the total occupation number of modes in the state $|g\rangle$. For a uniform source of finite size $R$ and ADM mass $M$, this amount can be estimated as

$$N_{\rm G} \sim \frac{M^2}{m_{\rm p}^2}\,\ln\!\left(\frac{R_\infty}{R}\right)\;, \tag{14}$$

where $R_\infty = k_0^{-1} \gg R$ is a IR cut-off, which cures the divergence coming from embedding a static (and eternal) field in an infinite space-time. In reference to that, it is interesting to point out that the dependence of $N_{\rm G}$ is weaker on $R$ than on the mass $M$, since

$$\frac{{\rm d}N_{\rm G}}{N_{\rm G}} \sim 2\,\frac{{\rm d}M}{M} - \frac{1}{\ln(R_\infty/R)}\,\frac{{\rm d}R}{R}\;, \tag{15}$$

and allows to recover the opening result $N_{\rm G} \sim M^2/m_{\rm p}^2$ when $R_\infty$ is arbitrarily large. We can now add the non-linearity described by Eq. (7), rewritten as

$$\triangle V_{(1)} = 2\,\frac{l_{\rm p}}{m_{\rm p}}\,\langle g|\left(\hat{\Phi}'\right)^2|g\rangle\;. \tag{16}$$

so that

$$2\,\frac{l_{\rm p}}{m_{\rm p}}\,\langle g|\left(\hat{\Phi}'\right)^2|g\rangle = J_g + J_0\;. \tag{17}$$

$J_0$ is a diverging contribution coming from the vacuum, which can be removed through normal ordering in the expectation value. By means of Eq. (12), one can immediately see that $J_g$ equals the classical expression, that is

$$J_g = 2\,\frac{l_{\rm p}}{m_{\rm p}}\,\langle g|\left(\hat{\Phi}'\right)^2|g\rangle = 2\left(V'_{(0)}\right)^2 , \tag{18}$$

for any current sourcing the scalar field. The coherent state $|g\rangle$ is therefore an appropriate basis for a perturbative analysis in Quantum Field Theory. In order to refine the result, one shall find a modified coherent state $|g'\rangle$, such that

$$\sqrt{\frac{l_{\rm p}}{m_{\rm p}}}\,\langle g'|\,\hat{\Phi}\,|g'\rangle \simeq V_{(0)} + q_\Phi\,V_{(1)} , \tag{19}$$

to first order in $q_\Phi$. As for the classical potential in Eq. (2), we can expand the quantum state $|g'\rangle$ as $|g'\rangle \simeq \mathcal{N}\left(|g\rangle + q_\Phi\,|\delta g\rangle\right)$, where $\mathcal{N}$ is a normalisation constant. After manipulating Eq. (19) with some cumbersome algebra [2], we succeed in relating the correction to any eigenvalue, $\delta g_k$, to the entire set of $g_p$'s, *i.e.*

$$\frac{\triangle\left(\mathrm{Re}\,\langle\delta g|\,\hat{\Phi}\,|g\rangle\right)}{\mathrm{Re}\,\langle\delta g|\,|g\rangle} = \triangle\,\langle g|\,\hat{\Phi}\,|g\rangle + \sqrt{\frac{l_{\rm p}}{m_{\rm p}}}\,\frac{\langle g|\left(\hat{\Phi}'\right)^2|g\rangle}{\mathrm{Re}\,\langle\delta g|\,|g\rangle} . \tag{20}$$

Of course, this Equation is very complicated and any attempt to deal with it without some sort of approximation is destined to fail. Therefore, we follow the argument outlined in Refs. [3, 15] and pretend that almost the whole set of toy gravitons belongs to one mode of wavelength $\lambda_{\rm G} \simeq R$ [8], so that one can finally identify [2]

$$\delta g_{\bar{k}} \simeq -l_{\rm p}\,\bar{k}^{3/2}\,\delta\bar{k}\,g_{\bar{k}}^2 \sim -l_{\rm p}\,\bar{k}^{5/2}\,g_{\bar{k}}^2 , \tag{21}$$

where $\delta\bar{k} \sim R^{-1}$. Considering *e.g.* the point-like source of Ref. [2], one obtains

$$\delta g_{\bar{k}} \sim -\frac{l_{\rm p}\,M_0^2}{m_{\rm p}^2\,\bar{k}^{1/2}} \sim \frac{l_{\rm p}\,M_0}{m_{\rm p}\,r_0}\,g_{\bar{k}} \sim \frac{R_{\rm H}}{r_0}\,g_{\bar{k}} , \tag{22}$$

with the help of a UV cut-off $r_0$, that removes the infinities coming from the vanishing spatial extension of the source. When the result falls within the range of validity of our approximation, that is $r_0 \ll R_{\rm H}$, we see that the perturbation correctly enjoys the relation $\delta g_{\bar{k}} \ll g_{\bar{k}}$, which is compatible with the classical result that we want to extend to the domain of quantum physics.

## 4. Conclusions and outlook

We have constructed an effective quantum theory for the gravitational potential sourced by a static matter distribution and up to first post-Newtonian order, by approximating the Einstein-Hilbert action in the weak field and non-relativistic regimes. The result [2] implies the maximal packing condition (2), which is a signature feature of the corpuscular BH model of Dvali and Gomez [11, 12]. Furthermore, we refined the expression (14) for the total number of soft quanta, $N_{\rm G}$, forming the self-sustained BEC and we showed that it is not only influenced by the total ADM mass of the black hole as in standard literature [13], but also encodes more information, albeit to a much weaker extent. In fact, the logarithmic dependence on the ratio $R/R_\infty$ becomes more and more negligible as the system approaches an ideal configuration, for which $R_\infty \to \infty$, while it is expected to give a non-trivial contribution in a dynamical situation. Moreover, we stated several times that our analysis looses accuracy for a source of size $R \lesssim R_{\rm H}$, as we have shown that the consistency relation $\delta g_k \ll g_k$ gets spoiled as one moves further away from it. Still, it would be interesting to investigate the case where the inequality is saturated, that is $R = R_{\rm H}$, since Eq. (2) arises precisely in this regime and allows the black hole to be self-sustained. The possible outcome may be able to quantify the departure of the present analysis from the standard post-Newtonian approximation of GR, possibly providing some glimpses of a full theory of Quantum Gravity. In addition to that, it is necessary to understand the role of matter pressure, which has been completely neglected so far, but may have important cosmological implications [37, 38].

## Acknowledgments

This proceeding is based on a series of papers in collaboration with R. Casadio, A. Giusti and M. Lenzi. The author is partially supported by the ERC Advanced Grant 339169 "Self-completion". This work has also been carried out in the framework of activities of the National Group of Mathematical Physics (GNFM, INdAM).

## References

[1] Weinberg S 1972 *Gravitation and Cosmology: Principles and Applications of the General Theory of Relativity* (New York: Wiley and Sons)
[2] Casadio R, Giugno A, Giusti A and Lenzi M 2017 *Phys. Rev.* D **96** no.4, 044010
[3] Casadio R, Giugno A and Giusti A 2016 *Phys. Lett.* B **763** 337
[4] Duff M J 1973 *Phys. Rev.* D **7** 2317
[5] Donoghue J F, Ivanov M M and Shkerin A 2017 *Preprint* arXiv:1702.00319 [hep-th]
[6] Casadio R, Giugno A and Giusti A 2017 *Gen. Rel. Grav.* **49** no.2, 32
[7] Casadio R, Giugno A and Micu O 2016 *Int. J. Mod. Phys.* D **25** no.02, 1630006
[8] Dvali G and Gomez C 2014 *JCAP* **01** 023
[9] Dvali G and Gomez C 2014 *Eur. Phys. J.* C **74** 2752
[10] Dvali G and Gomez C 2013 *Phys. Lett.* B **719** 419
[11] Dvali G and Gomez C 2012 *Phys. Lett.* B **716** 240
[12] Dvali G and Gomez C 2013 *Fortsch. Phys.* **61** 742
[13] Dvali G, Gomez C and Mukhanov S 2011 *Preprint* arXiv:1106.5894 [hep-ph]
[14] Flassig D, Pritzel A and Wintergerst N 2013 *Phys. Rev.* D **87** 084007
[15] Casadio R, Giugno A, Micu O and Orlandi A 2014 *Phys. Rev.* D **90** 084040
[16] Casadio R, Giugno A and Orlandi A 2015 *Phys. Rev.* D **91** 124069
[17] Casadio R, Giugno A, Micu O and Orlandi A 2015 *Entropy* **17** 6893
[18] Kühnel F 2014 *Phys. Rev.* D **90** 084024
[19] Kühnel F and Sundborg B 2014 *JHEP* **1412** 016
[20] Kühnel F and Sundborg B 2014 *Phys. Rev.* D **90** 064025
[21] Casadio R and Orlandi A 2013 *JHEP* **1308** 025
[22] Mück W and Pozzo G 2014 *JHEP* **1405** 128
[23] Casadio R, Giugno A, Giusti A and Micu O 2017 *Eur. Phys. J.* C **77** no.5, 322
[24] Ruffini R and Bonazzola S 1969 *Phys. Rev.* **187** 1767
[25] Colpi M, Shapiro S L and Wasserman I 1986 *Phys. Rev. Lett.* **57** 2485
[26] Membrado M, Abad J, Pacheco A F and Sanudo J 1989 *Phys. Rev.* D **40** 2736
[27] Nieuwenhuizen T M 2008 *Europhys. Lett.* **83** 10008
[28] Chavanis P-H and Harko T 2012 *Phys. Rev.* D **86** 064011
[29] Dvali G and Gußmann A 2016 *Nucl. Phys.* B **913** 1001
[30] Dvali G and Gußmann A 2017 *Phys. Lett.* B **768** (274-279
[31] Kühnel F and Sandstad M 2015 *Phys. Rev.* D **92** 124028
[32] Arnowitt R L, Deser S and Misner C W 1959 *Phys. Rev.* **116** 1322
[33] Bekenstein J D 1973 *Phys. Rev.* D **7** 2333
[34] Madsen M S 1988 *Class. Quant. Grav.* **5** 627
[35] Brown J D 1993 *Class. Quant. Grav.* **10** 1579
[36] Harko T 2010 Phys. Rev. D **81** 044021
[37] Casadio R, Giugno A and Giusti A 2017 *Preprint* arXiv:1708.09736 [gr-qc]
[38] Cadoni M, Casadio R, Giusti A, Mück W and Tuveri M 2017 *Preprint* arXiv:1707.09945 [gr-qc]

IOP Conf. Series: Journal of Physics: Conf. Series **942** (2017) 012013 doi:10.1088/1742-6596/942/1/012013

# Horizon Quantum Mechanics: spherically symmetric and rotating sources

**Andrea Giusti**

Department of Physics and Astronomy, University of Bologna, Via Irnerio 46, Bologna, Italy;
I.N.F.N., Sezione di Bologna, IS FLAG, viale B. Pichat 6/2, I-40127 Bologna, Italy;
Arnold Sommerfeld Center, Ludwig-Maximilians-Universität, Theresienstraße 37, 80333 München, Germany.

E-mail: agiusti@bo.infn.it

**Abstract.** In this paper we discuss some mathematical aspects of the horizon wave-function formalism, also known in the literature as horizon quantum mechanics. In particular, first we review the structure of both the global and local formalism for static spherically symmetric sources. Then, we present an extension of the global analysis for rotating black holes and we also point out some technical difficulties that arise while attempting the local analysis for non-spherically symmetric sources.

## 1. Introduction

Black hole physics, in the classical framework, is known to have an extremely fascinating and elegant mathematical formulation in terms of Lorentzian manifold. However, such a purely classical description leads to some major physical issues such as the appearance of *curvature singularities*, which is one of the main feature of black holes in general relativity [1, 2].

A first attempt at introducing some quantum mechanical effects in the context of Einstein's gravity was first proposed by Hawking [3], whose pioneering ultimately lead to the so called *semi-classical picture* of gravity. The latter basically consist in regarding black hole spacetimes as purely classical (*i.e.* geometrical) objects on which one can study the dynamics of some quantum fields. In other words, in this framework, black holes are understood as a purely classical background stage for the quantum matter fields. Needless to say that this quantum description of gravity does not tackle the problem of curvature singularities, but rather it leads to the rise of the so called *information paradox*.

A way to circumvent all these issues could potentially be obtained by changing our prospective on gravity. For example, from the point of view of a particle physicist, a black hole could be understood as a gravitationally bound state confining all possible signals within its horizon. Furthermore, if we look carefully at the physical features of gravity in the high curvature regime one could convince oneself that this picture is not that different from the one of quantum chromodynamics in the strongly coupled regime. An approach which supports this viewpoint is the so called corpuscular model of quantum black holes [4, 5, 6, 7, 8, 9, 10]. It is also worth remarking that this approach naturally solves the problem of singularity and the information paradox.

IOP Conf. Series: Journal of Physics: Conf. Series **942** (2017) 012013 doi:10.1088/1742-6596/942/1/012013

Here, we would like to offer an alternative to the traditional semi-classical gravity by presenting the so called horizon quantum mechanics [11, 12, 13] picture of gravity. Indeed, this formalism aims to a purely quantum mechanical description of trapping surfaces. In particular, the key idea is to consider the source for the black hole spacetime to be a quantum mechanical object from which we wish to infer what are the effects on the behaviour of the horizon due to the quantum nature of the source.

## 2. Static spherically symmetric systems

Let us start then with the description of both the global and local analysis for static and spherically symmetric black holes. In particular, here we present a very short review of the main results discussed in [14], therefore the interested reader should refer to this paper for further details.

A classical spherically symmetric spacetime can be described in terms of a line element of the form

$$\mathrm{d}s^2 = g_{tt}\,\mathrm{d}t^2 + g_{rr}\,\mathrm{d}r^2 + r^2\left(\mathrm{d}\theta^2 + \sin^2\theta\,\mathrm{d}\varphi^2\right) . \tag{1}$$

The Einstein's field equation for a static source then tell us that

$$g^{rr} = 1 - \frac{2\,G_{\mathrm{N}}\,E(r)}{r}\,, \qquad E(r) = 4\,\pi\int_0^r \rho(\bar{r})\,\bar{r}^2\,\mathrm{d}\bar{r}\,,$$

where $\rho(r)$ is the matter density of the source and $E(r)$ is the so called Misner-Sharp (or quasi-local) mass of the system[1]. If we now perform the study of the geodesic expansion one can easily see that the appearance of *trapping surfaces* corresponds to the condition $g^{rr} = 0$ and therefore to $r = 2\,G_{\mathrm{N}}\,E(r)$. Furthermore, it is worth remarking that if we take the limit for $r \to \infty$, then $E(r) \to M$, where $M$ is the Arnowitt-Deser-Misner (ADM) mass of the system which enters is the definition of the Schwarzschild radius.

If the source is purely quantum, the quantities that define the Misner-Sharp mass and ADM mass should become quantum variables themselves. Therefore, one expects the gravitational radius will undergo the same fate, leading to a sort of quantum fuzziness of Schwarzschild (or gravitational) radius of a localised quantum source.

Now, let us describe our system (source and geometry together) in terms of an Hilbert space $\mathcal{H}$ defined as

$$\mathcal{H} = \mathcal{H}_S \otimes \mathcal{H}_G \tag{2}$$

where $\mathcal{H}_S$ and $\mathcal{H}_G$ represent the Hilbert space for the source and for the geometry, respectively. The simplest way to reproduce the connection between the geometry of the spacetime and the source, at the quantum level, is obtained by considering an entangled state

$$\mathcal{H} \ni |\Psi\rangle = \sum_{\alpha\,,\beta} C(E_\alpha\,,\,R_{\mathrm{H}\,\beta})\;|E_\alpha\rangle\;|R_{\mathrm{H}\,\beta}\rangle \tag{3}$$

where the quantum states of the source correspond to the spectral decomposition of a certain hamiltonian operator $\widehat{H}$ (precisely, $\widehat{H}\otimes\mathbb{I}_G$) describing the source, while the quantum states of the geometry are labelled in terms of the spectral decomposition of a gravitational radius operator $\widehat{R}_{\mathrm{H}}$ (namely, $\mathbb{I}_S \otimes \widehat{R}_{\mathrm{H}}$). In other words, we have lifted the ADM mass and the Schwarzschild radius to operators on $\mathcal{H}$.

However, not all states of $\mathcal{H}$ can be seen as physical states for the system within this formalism. Indeed, the only physical states are the one that allow us to gain a Schwarzschild-like relation between the spectrum of $\widehat{H}$ and the one of $\widehat{R}_{\mathrm{H}}$, *i.e.* the quantum version of the relation

[1] Throughout the whole paper we will set the speed of light $c = 1$.

IOP Conf. Series: Journal of Physics: Conf. Series **942** (2017) 012013 doi:10.1088/1742-6596/942/1/012013

between the ADM mass and the gravitational radius. The physical configurations are therefore determined by the constraint

$$\left(\widehat{H} - \frac{1}{2\,G_{\mathrm{N}}}\,\widehat{R}_{\mathrm{H}}\right)\,|\Psi\rangle_{\mathbf{phys}} = 0\,, \tag{4}$$

that leads to $C(E_\alpha\,,\,R_{\mathrm{H}\,\beta}) = C(E_\alpha\,,\,2\,G_{\mathrm{N}}\,E_\alpha)\,\delta_{\alpha\beta}$.

Now, if we trace out the contribution of the gravitational degrees of freedom we get

$$|\Psi\rangle_{\mathrm{S}} = \sum_\gamma C_S(E_\gamma)\,|E_\gamma\rangle\,,\quad C_S(E_\gamma) \equiv C(E_\gamma\,,\,2\,G_{\mathrm{N}}\,E_\gamma) \tag{5}$$

which represents the spectral decomposition of the source, whereas if we trace away the contribution of the source we get

$$|\Psi\rangle_{\mathrm{H}} = \sum_\gamma C_S(R_{\mathrm{H}\,\gamma}/2\,G_{\mathrm{N}})\,|R_{\mathrm{H}\,\gamma}\rangle\,, \tag{6}$$

which is known as the **horizon wave-function** (HWF) and represents the fundamental ingredient of the horizon quantum mechanics.

This object allows us to provide a *probabilistic definition of black hole*. Indeed, if we assume a continuous spectrum for our observables, the HWF in the gravitational radius representation will be given by

$$\Psi_{\mathrm{H}}(R_{\mathrm{H}}) := \langle R_{\mathrm{H}}\,|\,\Psi_{\mathrm{H}}\rangle = \mathcal{N}_{\mathrm{H}}\,C_S(R_{\mathrm{H}\,\gamma}/2\,G_{\mathrm{N}})\,, \tag{7}$$

where $\mathcal{N}_{\mathrm{H}}$ is just a normalisation constant. Then, we can build a *probability density function* that one would detect a gravitational radius of size $R_{\mathrm{H}}$ associated with the particle in the quantum state $|\Psi\rangle_{\mathrm{S}}$, *i.e.*

$$\mathcal{P}_{\mathrm{H}}(r_{\mathrm{H}}) := 4\,\pi\,r_{\mathrm{H}}^2\,|\Psi_{\mathrm{H}}(R_{\mathrm{H}})|^2\,. \tag{8}$$

Hence, we can now define the *probability density for a quantum particle to be a black hole* as the conditional probability density that the particle lies inside its own gravitational radius $R_{\mathrm{H}}$, *i.e.*

$$\mathcal{P}_<(r < r_{\mathrm{H}}) := \mathcal{P}_{\mathrm{S}}(r < R_{\mathrm{H}})\,\mathcal{P}_{\mathrm{H}}(R_{\mathrm{H}})\,, \tag{9}$$

where

$$\mathcal{P}_{\mathrm{S}}(r < r_{\mathrm{H}}) = 4\,\pi\int_0^{R_{\mathrm{H}}} \bar{r}^2\,\Psi_{\mathrm{S}}(\bar{r})\,\mathrm{d}\bar{r}\,.$$

Then, the *probability for a quantum source to be a black hole* is just given by integrating (9) over all possible values of $R_{\mathrm{H}} \in (0,\infty)$.

This formalism also allows for a local description of the gravitational radius, *i.e.* for localised energy eigenmodes and correspond- ingly discrete energy quantum numbers. To see that, let us restrict our attention to what happens within a sphere of radius $r$. In this case, the role the ADM mass is now played by the Misner-Sharp mass $E(r)$, therefore the new hamiltonian operator will now be dependent on the radius, *i.e.* $E(r) \mapsto \widehat{H}(r)$. It is easy to prove that the spectrum of $\widehat{H}(r)$ will be given by

$$\widehat{H}(r) = \sum_\alpha P_\alpha(r)\,E_\alpha\,|E_\alpha\rangle\,\langle E_\alpha|\,,\quad P_\alpha(r) := 4\,\pi\int_0^r \bar{r}^2\,\Psi_{E_\alpha}(\bar{r})\,\mathrm{d}\bar{r}\,. \tag{10}$$

Notice that $\lim_{r\to\infty} P_\alpha(r) = 1$ only for $\Psi_{E_\alpha} \in L_2(0,\infty)$, *i.e.* for localised energy eigenmodes. This restriction was not a requirement in the global case, since the norm of these modes never entered explicitly in that calculation.

IOP Conf. Series: Journal of Physics: Conf. Series **942** (2017) 012013 doi:10.1088/1742-6596/942/1/012013

If we proceed as above, we can ultimately provide a local (quantum-mechanical) characterisation of trapping surfaces, specifically one finds that

$$\langle r_{\rm H} |\, r_{\rm H}(r)\, | r_{\rm H} \rangle = r \,, \qquad r_{{\rm H}\,\alpha}(r) := P_\alpha(r)\, R_{{\rm H}\,\alpha} \,, \tag{11}$$

which is the local quantum version the condition $g^{rr} = 0$ in terms of the Misner-Sharp mass.

## 3. Rotating sources: global formalism

The prototype for a rotating black hole, in general relativity, is given by the Kerr solution of the vacuum Einstein's field equations, *i.e.*

$$\mathrm{d}s^2 \;=\; -\left(1 - \frac{r_s\, r}{\Sigma}\right) \mathrm{d}t^2 + \frac{\Sigma}{\Delta}\, \mathrm{d}r^2 + \Sigma\, \mathrm{d}\theta^2 \tag{12}$$

$$+ \; \left(r^2 + a^2 + \frac{r_s\, r\, a^2}{\Sigma} \sin^2\theta\right) \sin^2\theta\, \mathrm{d}\phi^2 - \frac{2\, r_s\, r\, a\, \sin^2\theta}{\Sigma}\, \mathrm{d}t\, \mathrm{d}\phi \tag{13}$$

where $M$ is the ADM mass, $r_s = 2\,G_{\rm N}\, M$, $a = J/M$ with $J$ the angular momentum of the system, $\Sigma = r^2 + a^2 \cos^2\theta$ and $\Delta = r^2 - r_s\, r + a^2$.

For this spacetime, the horizons are given by the condition $\Delta = 0$, which in turns leads to

$$R_{\rm H}^{(\pm)} = G_{\rm N} \left(M \pm \sqrt{M^2 - \frac{J^2}{M^2}}\right) \tag{14}$$

As stated in the Carter-Robinson uniqueness theorem, a vacuum black hole solution in an stationary, axisymmetric and asymptotically flat spacetime belongs to the Kerr family of solutions vacuum Einstein's field equations, which in turns is parametrised by two real parameters $(M, J)$, *i.e.* the ADM mass and the total angular momentum.

Therefore, if we want to perform an analysis in the framework of the horizon quantum mechanics [15], we have to find a way to promote the two parameters $(M, J)$ to quantum mechanical observables. However, this should be done with a bit of care due to the fact that the notion of angular momentum in quantum mechanics is extremely different from the classical one.

So, let us assume the existence of a complete set of commuting operators connected with quantum source, *i.e.* $\{\widehat{H},\, \widehat{J}^2,\, \widehat{J}_z\}$, respectively the Hamiltonian, the total angular momentum squared and the angular momentum along the axis of rotation.

Then, we can just promote the two parameters $(M, J)$ to operators on $\mathcal{H}$ as follows

$$M \;\mapsto\; \widehat{H} = \sum_{a,j,m} E_{a\,j}\; |a\, j\, m\rangle\, \langle a\, j\, m| \;, \tag{15}$$

$$J^2 \;\mapsto\; \widehat{J}^2 = m_{\rm p}^4 \sum_{a,j,m} j(j+1)\; |a\, j\, m\rangle\, \langle a\, j\, m| \equiv \sum_{a,j,m} \lambda_j\; |a\, j\, m\rangle\, \langle a\, j\, m| \,, \tag{16}$$

$$\widehat{J}_z = m_{\rm p}^2 \sum_{a,j,m} m\; |a\, j\, m\rangle\, \langle a\, j\, m| \equiv \sum_{a,j,m} \xi_m\; |a\, j\, m\rangle\, \langle a\, j\, m| \;. \tag{17}$$

where $j \in \mathbb{N}_0/2$, $m \in \mathbb{Z}/2$, with $|m| \le j$, and $a \in \mathcal{I}$, with $\mathcal{I}$ a *discrete* set of labels that can be either finite or infinite.

In analogy with the previous section, the quantum state for the whole system (source plus geometry) can be expressed as a triple entangled state, *i.e.*

$$|\Psi\rangle = \sum_{a,j,m} \sum_{\alpha,\beta} C(E_{a\,j}, \lambda_j, \xi_m, R_{{\rm H}\alpha}^{(+)}, R_{{\rm H}\beta}^{(-)})\; |a\, j\, m\rangle\, |\alpha\rangle_+\, |\beta\rangle_- \tag{18}$$

IOP Conf. Series: Journal of Physics: Conf. Series **942** (2017) 012013 doi:10.1088/1742-6596/942/1/012013

where $|\alpha\rangle_\pm$ are the eigenstates of $\widehat{R}_{\rm H}^{(\pm)}$.

Now, following the same line of thought as before, we can select the physical states for our system by imposing two constraints on $|\Psi\rangle \in \mathcal{H}$, namely

$$\left(\widehat{R}_{\rm H}^{(+)} - \widehat{\mathcal{O}}^{+}\right)|\Psi\rangle_{\rm phys} = 0\,, \quad \left(\widehat{R}_{\rm H}^{(-)} - \widehat{\mathcal{O}}^{-}\right)|\Psi\rangle_{\rm phys} = 0\,, \tag{19}$$

with

$$\widehat{\mathcal{O}}^{\pm} := \widehat{H} \pm \left(\widehat{H}^2 - \widehat{J}^2\,\widehat{H}^{-2}\right)^{1/2}\,,$$

provided that $\widehat{\mathcal{O}}^{\pm}$ is positively semi-definite.

Then, following the same procedure as in the previous section, we can now compute the (unnormalized) HWF for each horizon and, consequently, the black hole probability.

It is worth remarking that, due to the lack of a proper generalization of the Misner-Sharp mass for non-spherically symmetric systems, the local formalism for rotating sources is still an open problem.

## 4. Conclusions and outlook

We have been able to provide a precise mathematical formulation of both the global and local horizon quantum mechanics for static spherically symmetric quantum sources. Specifically, we have introduced the notion of probabilistic definition of black hole and we have provided a complete characterisation of trapping surfaces at the quantum level.

Furthermore, we have also been able to extend the global formalism to the case of rotating sources described in terms of a metric of the Kerr family. However, a complete quantum mechanical local analysis requires a generalised notion of Misner-Sharp mass for axisymmetric system which is, however, still an open problem in classical general relativity.

Despite the many works talking the implementation of the horizon quantum mechanics for classical solutions of Einstein's field equations, very few papers have been dealing with metric describing non-singular spacetimes. Therefore, it would be interesting to see an application of this formalism to the Hayward spacetime [16, 17], and its rotating counterpart (see *e.g.* [18, 19]).

## References

[1] Wald R, General relativity. University of Chicago press, 2010.
[2] Straumann N, General relativity and relativistic astrophysics. Springer, 2012.
[3] Hawking S W 1975 *Commun. Math. Phys.* **43** 199
[4] Dvali G and Gomez C 2013 *Phys. Lett.* B **719** 419
[5] Dvali G and Gomez C 2012 *Phys. Lett.* B **716** 240
[6] Casadio R, Giugno A and Giusti A 2016 *Phys. Lett.* B **763** 337
[7] Casadio R, Giugno A and Orlandi A 2015 *Phys. Rev.* D **91** 124069
[8] Casadio R, Giugno A, Micu O and Orlandi A 2014 *Phys. Rev.* D **90** 084040
[9] Casadio R, Giugno A, Giusti A and Lenzi M 2017 *Phys. Rev.* D **96** no.4, 044010
[10] Casadio R, Giugno A and Giusti A 2017 *Preprint* arXiv:1708.09736 [gr-qc]
[11] Casadio R and Scardigli F 2014 *Eur. Phys. J. C* **74** no.1, 2685
[12] Casadio R, Giugno A and Micu O 2016 *Int. J. Mod. Phys.* D **25** no.02, 1630006
[13] Casadio R, Giugno A, Micu O and Orlandi A 2015 *Entropy* **17** 6893
[14] Casadio R, Giugno A and Giusti A 2017 *Gen. Rel. Grav.* **49**, no. 2, 32
[15] Casadio R, Giugno A, Giusti A and Micu O 2017 *Eur. Phys. J.* C **77** no.5, 322
[16] Hayward S A 2006 *Phys.Rev.Lett.* **96**, 031103
[17] Giugno A, Giusti A and Helou A 2017, in preparation
[18] Bambi C and Modesto L 2013 *Phys. Lett. B* **721**, 329
[19] De Lorenzo T, Giusti A and Speziale S 2016 *Gen. Rel. Grav.* **48** no.3, 31

IOP Conf. Series: Journal of Physics: Conf. Series **942** (2017) 012014 doi:10.1088/1742-6596/942/1/012014

# Quasinormal Modes of Static Modified Gravity (MOG) Black Holes

**Luciano Manfredi**[1], **Jonas Mureika**[1], **John Moffat**[2]

1. Department of Physics, Loyola Marymount University, Los Angeles, CA USA
2. Perimeter Institute for Theoretical Physics, Waterloo, ON CA/N

E-mail: lmanfred@lion.lmu.edu; jmureika@lmu.edu; jmoffat@perimeterinstitute.ca

**Abstract.** Using an asymptotic iteration method, we calculate the gravitational and electromagnetic quasinormal mode (QNM) perturbations for a static neutral black hole described by a Scalar-Tensor-Vector Modified Gravity framework (STVG-MOG). We show that the first few harmonic modes differ from their general relativistic (GR) equivalent for a Schwarzschild black hole. Specifically, the real and imaginary components of the QNM frequencies are smaller for STVG-MOG than for GR. We posit that the differences are sufficiently large to potentially be observed in present and future black hole binary merger gravitational waveforms.

## 1. Introduction

The landmark detection of gravitational waves from binary black hole coalescence by LIGO [1, 2, 3, 4] has ushered in a new age in astronomy and black hole physics. Due to the unique characteristics of the waveforms in a given spacetime, this data present us with a new test of the underlying theory of gravity, be it General Relativity [5] or some other alternative theory [6, 7, 8, 9].

The observed merger signals are divided into three phases – the inspiral, merger, and ringdown, – each of which identifies key characteristics of the black holes (size, mass, spin, *etc...*). While much attention is paid to the "loud" inspiral phase and the direct information it can disclose about the black holes, the more subtle ringdown phase can provide addition insight into the underlying spacetime structure. During this period, the final black hole settles to an unperturbed state through the emission of additional gravitational radiation, through damped oscillations known as quasinormal modes (QNMs) [10]. These QNMs are likely to be visible in present or future LIGO data [11, 12, 13], and most definitely by LISA [14]. An array of model-dependent data from QNMs can help to identify the background spacetime structure, including the frequencies themselves, QNM onset time [15], as well as higher-order modes [16].

## 2. STVG-MOG Overview

In this proceedings contribution, we summarize our analysis of the ringdown phase of black hole mergers in a scalar-vector-tensor theory of gravitation also known as SVTG-MOG [17, 18, 19, 20, 21]. The theory in question is best thought of as a generalization of Einstein gravity in which the metric theory of gravitation is supplemented by two auxilary fields, *i.e.* a

IOP Conf. Series: Journal of Physics: Conf. Series **942** (2017) 012014 doi:10.1088/1742-6596/942/1/012014

scalar and vector graviton. These are described by the equations

$$G_{\mu\nu} - \Lambda g_{\mu\nu} + Q_{\mu\nu} = -8\pi G T_{\mu\nu}, \tag{1}$$

$$\frac{1}{\sqrt{-g}}\partial_\nu(\sqrt{-g}B^{\mu\nu}) + \mu^2\phi^\mu = -J^\mu, \tag{2}$$

$$\partial_\sigma B_{\mu\nu} + \partial_\mu B_{\nu\sigma} + \partial_\nu B_{\sigma\mu} = 0, \tag{3}$$

with the parameters $\Box G = K(x)$ and $\Box\mu = L(x)$. The advantages of this framework include the elimination of dark matter in galaxy rotation curves and clusters [22, 23, 24], as well as structure formation in the early universe [25, 26, 27].

The additional tensor $Q_{\mu\nu}$ introduced into Einstein's equations (1) is defined as

$$Q_{\mu\nu} = \frac{2}{G^2}(\partial^\alpha G\partial_\alpha G g_{\mu\nu} - \partial_\mu G\partial_\nu G) - \frac{1}{G}(\Box G g_{\mu\nu} - \nabla_\mu\partial_\nu G). \tag{4}$$

where

$$K(x) = \frac{3}{G}\left(\frac{1}{2}\partial^\alpha G\partial_\alpha G - V(G)\right) - \frac{3}{G}\partial^\alpha G\partial_\alpha G + \frac{G}{\mu^2}\left(\frac{1}{2}\partial^\alpha\mu\partial_\alpha\mu - V(\mu)\right) + \frac{3G^2}{16\pi}\Box\left(\frac{1}{G}\right), \tag{5}$$

and

$$L(x) = -\left(\frac{1}{G}\partial^\alpha G\partial_\alpha\mu + \frac{2}{\mu}\partial^\alpha\mu\partial_\alpha\mu - \mu^2 G\frac{\partial V(\phi_\mu))}{\partial\mu}\right). \tag{6}$$

The total energy-momentum tensor is $T_{\mu\nu} = T^M_{\mu\nu} + T^\phi_{\mu\nu} + T^G_{\mu\nu} + T^\mu_{\mu\nu}$, where $T^M_{\mu\nu}$ is the energy-momentum tensor for ordinary matter, and those of the additional fields are

$$T^{(\phi)}_{\mu\nu} = -\frac{1}{4\pi}\left[B_\mu{}^\alpha B_{\nu\alpha} - g_{\mu\nu}\left(\frac{1}{4}B^{\rho\alpha}B_{\rho\alpha} + V(\phi_\mu)\right) + 2\frac{\partial V(\phi_\mu)}{\partial g^{\mu\nu}}\right], \tag{7}$$

$$T^{(G)}_{\mu\nu} = -\frac{1}{4\pi G^3}\left(\partial_\mu G\partial_\nu G - \frac{1}{2}g_{\mu\nu}\partial_\alpha G\partial^\alpha G\right) \;,\; T^{(\mu)}_{\mu\nu} = -\frac{1}{4\pi G\mu^2}\left(\partial_\mu\mu\partial_\nu\mu - \frac{1}{2}g_{\mu\nu}\partial_\alpha\mu\partial^\alpha\mu\right). \tag{8}$$

It is possible to obtain a generalized "Schwarzschild-MOG" solution of these field equations by demanding that $G = G_N(1+\alpha)$ and $Q_g = \sqrt{\alpha G_N}M$ be constant. Since the $\phi_\mu$ field particle mass can be fit as $m_\phi \sim 10^{-28}$ eV in the present universe [22, 23], we can ignore this contribution and rewrite the field equations and modified $T^{(\phi)}{}_\mu{}^\nu$ as follows:

$$R_{\mu\nu} = -8\pi G T^\phi_{\mu\nu}, \tag{9}$$

$$\frac{1}{\sqrt{-g}}\partial_\nu(\sqrt{-g}B^{\mu\nu}) = 0, \tag{10}$$

$$\partial_\sigma B_{\mu\nu} + \partial_\mu B_{\nu\sigma} + \partial_\nu B_{\sigma\mu} = 0. \tag{11}$$

$$T^{(\phi)}{}_\mu{}^\nu = -\frac{1}{4\pi}(B_{\mu\alpha}B^{\nu\alpha} - \frac{1}{4}\delta_\mu{}^\nu B^{\alpha\beta}B_{\alpha\beta}). \tag{12}$$

The solution to the above system yields the metric

$$ds^2 = f(r)dt^2 - f(r)^{-1}dr^2 - r^2 d\Omega^2 \;,\; f(r) = \left(1 - \frac{2G_N(1+\alpha)M}{r} + \frac{\alpha(1+\alpha)G_N^2M^2}{r^2}\right) \tag{13}$$

Algebraically, this is identical in form to the Riessner-Nordström solution, but with the electric charge term now dependent on the mass.

## 3. Calculating Quasinormal Modes via the Asymptotic Iteration Method

New insights to black hole physics can be obtained through the study of these objects with surrounding fields. An understanding of QNMs, characteristic to the background spacetime, can then serve as a unique fingerprint when searching for the existence of black holes. These resonant modes present a real part that describes the actual frequency, and an imaginary part that dictates the damping. Gravitational perturbations appear to be of special interest since they directly identify the black hole to their gravitational radiation.

In the following we shall use the definition that QNMs are defined as solutions of the perturbed field equations with boundary conditions:

$$\psi(x) \longrightarrow \begin{cases} e^{-i\omega x} & x \to -\infty \\ e^{i\omega x} & x \to \infty \end{cases} \tag{14}$$

where the positive and negative solutions correspond to ingoing and outgoing waves respectively.

A number of methods have been developed to calculate QNMs for black holes, but one of the more robust semianalytic techniques is the asymptotic iteration method (AIM) [28]. This will be the algorithm of choice outlined below.

We begin by defining a second-order differential equation of the form

$$\chi'' = \lambda_0(x)\chi' + s_0(x)\chi, \tag{15}$$

The functions $\lambda_0(x)$ and $s_0(x)$ are smooth, so we can differentiate to get

$$\chi''' = \lambda_1(x)\chi' + s_1(x)\chi, \tag{16}$$

whose coefficients now satisfy $\lambda_1(x) = \lambda_0' + s_0 + \lambda_0^2$ and $s_1(x) = s_0' + s_0\lambda_0$. Iterating the differentiation approach we arrive to

$$\chi^{(n+2)} = \lambda_n(x)\chi' + s_n(x)\chi, \tag{17}$$

where the coefficients satisfy

$$\lambda_n(x) = \lambda_{n-1}' + s_{n-1} + \lambda_0\lambda_{n-1}, \qquad s_n(x) = s_{n-1}' + s_0\lambda_{n-1}. \tag{18}$$

For $n$ large enough, the AIM feature is introduced

$$\frac{s_n(x)}{\lambda_n(x)} = \frac{s_{n-1}(x)}{\lambda_{n-1}(x)} \equiv \beta(x), \tag{19}$$

and the QNMs arise from a "quantization condition" marking the end of the iterations

$$\delta_n = s_n\lambda_{n-1} - s_{n-1}\lambda_n = 0. \tag{20}$$

Outside the horizon, we describe the perturbations of the geometry as even- and odd-parity oscillations of a Schrödinger-like equation, *i.e.*

$$\left(\frac{d^2}{dx^2} - \rho^2 - V_i^{(\pm)}\right) Z_i^{(\pm)} = 0, \tag{21}$$

with the even and odd modes denoted by + and −, respectively [30, 31, 32]:

$$V_i^{(-)}(r) = \tfrac{\Delta}{r^5}\left(Ar - q_j + \tfrac{4Q_g^2}{r}\right) \tag{22}$$

$$V_i^{(+)}(r) = V_i^{(-)}(r) + 2q_j\tfrac{d}{dx}\left(\tfrac{\Delta}{r^2[(l-1)(l+2)r+q_j]}\right). \tag{23}$$

IOP Conf. Series: Journal of Physics: Conf. Series **942** (2017) 012014 doi:10.1088/1742-6596/942/1/012014

When $i = j = (1,2)(i \neq j)$, we can write

$$\frac{dr}{dx} = \frac{\Delta}{r^2}, \tag{24}$$

$$\Delta = (r - r_+)(r - r_-) = \frac{1}{4}[4r^2 - 4r(1+\alpha) + \alpha(1+\alpha)], \tag{25}$$

$$A = l(l+1), \tag{26}$$

$$q_1 = \frac{1}{2}\left[3 + \sqrt{9 + 16Q_g^2(l-1)(l+2)}\right], \tag{27}$$

$$q_2 = \frac{1}{2}\left[3 - \sqrt{9 + 16Q_g^2(l-1)(l+2)}\right], \tag{28}$$

Under an appropriate change of variables, the AIM formalism can be expressed as

$$\chi_{Z_{i,\xi\xi}} = \lambda_{Z_i}(\xi)\chi_{Z_{i,\xi}} + s_{Z_i}(\xi)\chi_{Z_i}, \tag{29}$$

where

$$\lambda_{Z_i}(\xi) = \frac{\alpha(\alpha+1) + 4\Delta}{2\Delta(1-\xi)} - \frac{r_+(\alpha + 2\Gamma_Z\Delta + 1)}{\Delta(\xi-1)^2}, \tag{30}$$

$$s_{Z_i}(\xi) = \frac{r_+^6\left(\rho^2 + V_i^{(-)}\right)}{\Delta^2(\xi-1)^8} - \frac{\Gamma_Z r_+^2(1+\alpha+\Delta\Gamma_Z)}{\Delta(\xi-1)^4} - \frac{\alpha(\alpha+1)\Gamma_Z r_+}{2\Delta(\xi-1)^3} - \frac{r_+\Gamma_{Z,\xi}}{(\xi-1)^2}, \tag{31}$$

$$\Gamma_Z = -\rho - \frac{1-\xi}{r_+} + \frac{\left[(1-\rho)(r_+ - r_-) - \rho r_+^2\right](1-\xi)}{(r_+ - r_-)\left[r_+ - r_-(1-\xi)\right]} + \frac{\rho r_+(1-\xi)}{(r_+ - r_-)\xi}, \tag{32}$$

$$V_i^{(-)} = \Delta\frac{(1-\xi)^5}{r_+^5}\left[\frac{Ar_+}{1-\xi} - q_j + \frac{4Q_g^2(1-\xi)}{r_+}\right], \tag{33}$$

$$\Delta = \frac{r_+\xi\left[r_+ - r_-(1-\xi)\right]}{(1-\xi)^2} = \frac{(1+\alpha)\xi\left(2 + 2\sqrt{1+\alpha} + \alpha\xi\right)}{4(\xi-1)^2}, \tag{34}$$

$$r_\pm = \frac{1}{2}\left[1 + \alpha \pm (1+\alpha)^{1/2}\right], \tag{35}$$

$$Q_g = \frac{\sqrt{\alpha}}{2}. \tag{36}$$

Tables 1 and 2 show the real and complex QNM frequencies for the first few modes. The black hole mass has been set to unity for easy comparison with existing QNM for other gravitational theories. It is clearly shown that for $\alpha = 0$, the QNMs of $V_{i=2}^{(-)}$ in Table 2 reduce to the purely gravitational Schwarzschild case, while the QNMs of $V_{i=1}^{(-)}$ in Table 1 reduce to the purely electromagnetic case. It is important to note that for $n \gg l$ as in Table 1 for $l = 1$ and $n = 3$, a higher number of iterations is needed to find stable solutions in the semianalytical approach.

## 4. Conclusions

Quasinormal modes in the ringdown phase of black hole mergers described by STVG (MOG) theory were determined for both gravitational and electromagnetic perturbations. For increasing model parameter $\alpha > 0$, the magnitude of both the real and imaginary components of the frequencies decreased in the gravitational $l = 2$, $l = 3$ and $l = 4$ modes, as well as for the $l = 1$, $l = 2$ and $l = 3$ electromagnetic modes. This difference between MOG and GR should be detectable for sufficiently sensitive frequency determinations in present and future black hole binary coalescence data.

IOP Conf. Series: Journal of Physics: Conf. Series **942** (2017) 012014 doi:10.1088/1742-6596/942/1/012014

**Table 1.** QNMs accurate to 4 decimal places for $M = 1$ scaled MOG electromagnetic perturbations $V^{(-)}_{i=1}$ for $l = 1$, $l = 2$ and $l = 3$ modes.

| l | n | $\alpha = 0$ | $\alpha = 1$ | $\alpha = 4$ | $\alpha = 9$ |
|---|---|---|---|---|---|
| 1 | 0 | 0.2483 - 0.09249i | 0.1448 - 0.04805i | 0.06343 - 0.01881i | 0.03268 - 0.009084i |
| | 1 | 0.2145 - 0.2937i | 0.1308 - 0.1506i | 0.05882 - 0.05828i | 0.03038 - 0.02796i |
| | 2 | 0.1748 - 0.5252i | 0.1135 - 0.2654i | 0.05258 - 0.1014i | 0.02675 - 0.04833i |
| | 3 | 0.1462 - 0.7719i | 0.1090 - 0.3866i | 0.05036 - 0.1494i | 0.02434 - 0.07008i |
| 2 | 0 | 0.4576 - 0.09500i | 0.2651 - 0.04917i | 0.1164 - 0.01930i | 0.06000 - 0.009351i |
| | 1 | 0.4365 - 0.2907i | 0.2565 - 0.1498i | 0.1136 - 0.05854i | 0.05861 - 0.02830i |
| | 2 | 0.4012 - 0.5016i | 0.2420 - 0.2563i | 0.1087 - 0.09948i | 0.05607 - 0.04791i |
| | 3 | 0.3626 - 0.7302i | 0.2257 - 0.3699i | 0.1028 - 0.1425i | 0.05272 - 0.06840i |
| 3 | 0 | 0.6569 - 0.09562i | 0.3771 - 0.04933i | 0.1648 - 0.01936i | 0.08493 - 0.009395i |
| | 1 | 0.6417 - 0.2897i | 0.3709 - 0.1492i | 0.1627 - 0.05842i | 0.08391 - 0.02831i |
| | 2 | 0.6138 - 0.4921i | 0.3594 - 0.2522i | 0.1589 - 0.09841i | 0.08198 - 0.04760i |
| | 3 | 0.5779 - 0.7063i | 0.3446 - 0.3600i | 0.1537 - 0.1398i | 0.07927 - 0.06743i |

**Table 2.** QNMs accurate to 4 decimal places for $M = 1$ scaled MOG gravitational perturbations $V^{(-)}_{i=2}$ for $l = 2$, $l = 3$ and $l = 4$ modes.

| l | n | $\alpha = 0$ | $\alpha = 1$ | $\alpha = 4$ | $\alpha = 9$ |
|---|---|---|---|---|---|
| 2 | 0 | 0.3737 - 0.0890i | 0.2220 - 0.04650i | 0.1021 - 0.01867i | 0.05431 - 0.009171i |
| | 1 | 0.3467 - 0.2739i | 0.2115 - 0.1423i | 0.09872 - 0.05678i | 0.05270 - 0.02781i |
| | 2 | 0.3011 - 0.4783i | 0.1937 - 0.2457i | 0.09283 - 0.09696i | 0.04974 - 0.04725i |
| | 3 | 0.2515 - 0.7051i | 0.1742 - 0.3579i | 0.08582 - 0.1397i | 0.04584 - 0.06776i |
| 3 | 0 | 0.5994 - 0.0927i | 0.3353 - 0.04758i | 0.1496 - 0.0189i | 0.07875 - 0.009267i |
| | 1 | 0.5826 - 0.2813i | 0.3281 - 0.1441i | 0.1472 - 0.0571i | 0.07761 - 0.02795i |
| | 2 | 0.5517 - 0.4791i | 0.3149 - 0.2444i | 0.1428 - 0.0964i | 0.07543 - 0.04706i |
| | 3 | 0.5120 - 0.6903i | 0.2979 - 0.3503i | 0.1368 - 0.1373i | 0.07238 - 0.06680i |
| 4 | 0 | 0.8092 - 0.0942i | 0.4452 - 0.04804i | 0.1965 - 0.01903i | 0.1030 - 0.009311i |
| | 1 | 0.7966 - 0.2843i | 0.4398 - 0.1449i | 0.1947 - 0.05731i | 0.1021 - 0.02802i |
| | 2 | 0.7727 - 0.4799i | 0.4294 - 0.2441i | 0.1912 - 0.09625i | 0.1004 - 0.04699i |
| | 3 | 0.7398 - 0.6839i | 0.4151 - 0.3468i | 0.1863 - 0.1362i | 0.09796 - 0.06636i |

## Acknowledgments

L. Manfredi thanks the organizers of the KSM2017 for their generous hospitality during the meeting. LM and J. Mureika also thank thank the Perimeter Institute for Theoretical Physics for its generous hospitality during the initial stages of this research. J. Moffat thanks Luis Lehner and Martin Green for helpful discussions. This work was supported in part by by a Summer Undergraduate Research Fellowship and Continuing Faculty Grant from the Frank R. Seaver College of Science and Engineering of Loyola Marymount University and the Perimeter Institute for Theoretical Physics. Research at Perimeter Institute is supported by the Government of Canada through the Department of Innovation, Science and Economic Development Canada and by the Province of Ontario through the Ministry of Research, Innovation and Science.

## References

[1] Abbott B P *et al.* [LIGO Scientific and Virgo Collaborations], Phys. Rev. Lett. **116**, 061102 (2016) *Preprint* 1602.03837 [gr-qc].

[2] Abbott B P *et al.* [LIGO Scientific and Virgo Collaborations], Phys. Rev. Lett. **116**, 241103 (2016) *Preprint* 1606.04855 [gr-qc].

[3] Abbott B P *et al.* [LIGO Scientific and VIRGO Collaborations], Phys. Rev. Lett. **118**, 221101 (2017) *Preprint* 1706.01812 [gr-qc].
[4] Abbott B P *et al.* [LIGO Scientific and Virgo Collaborations], Phys. Rev. Lett. **119** (2017) 141101 *Preprint* 1709.09660 [gr-qc].
[5] Abbott B P *et al.* [LIGO Scientific and Virgo Collaborations], Phys. Rev. Lett. **116**, 221101 (2016) *Preprint* 1602.03841 [gr-qc].
[6] Konoplya R A and Zhidenko A, Rev. Mod. Phys. **83**, 793 (2011) *Preprint* 1102.4014 [gr-qc].
[7] Sibandze D B, Goswami R, Maharaj S D and Dunsby P K S, "Echoes from the black holes: Evidence of higher order corrections to General Relativity in strong gravity regime," *Preprint* 17020.4926 [qr-qc].
[8] Lasky P D and Doneva D D, Phys. Rev. **D82**, 124068 (2010) *Preprint* gr-qc/10110747.
[9] Bhattacharyya S and Shankaranarayanan S, "Quasinormal modes as a distinguisher between general relativity and f(R) gravity," *Preprint* gr-qc/170407044.
[10] Kokkotas K D and Schmidt B G, Living Rev. Rel. **2**, 2 (1999) *Preprint* gr-qc/9909058.
[11] Goggin L M, "A Search For Gravitational Waves from Perturbed Black Hole Ringdowns in LIGO Data," *Preprint* gr-qc/09082085.
[12] Abbott B P*et al.* [LIGO Scientific Collaboration], Phys. Rev. **D80**, 062001 (2009) *Preprint* gr-qc/09051654.
[13] Aasi J *et al.* [LIGO Scientific and VIRGO Collaborations], Phys. Rev. **D89**, no. 10, 102006 (2014) *Preprint* gr-qc/14035306.
[14] Berti E, Cardoso V and Will C M, Phys. Rev. **D73**, 064030 (2006) *Preprint* gr-qc//0512160.
[15] Sakai K, Oohara K I, Nakano H, Kaneyama M and Takahashi H, "Estimation of starting times of quasi-normal modes in ringdown gravitational waves with the Hilbert-Huang transform," *Preprint* gr-qc/170504107.
[16] London L, Shoemaker D and Healy J, Phys. Rev. **D90**, no. 12, 124032 (2014) Erratum: [Phys. Rev. **D94**, no. 6, 069902 (2016)] *Preprint* gr-qc/14043197
[17] Moffat J W, JCAP 0603, 004 (2006). *Preprint* gr-qc/0506021.
[18] Moffat J W, Phys. Lett. **B763**, 427 (2016), *Preprint* gr-qc/160305225.
[19] Moffat J W, Eur. Phys. J. C**75**, 175 (2015), *Preprint* gr-qc/14125429.
[20] Moffat J W, Eur. Phys. J. C **75**, 130 (2015), *Preprint* gr-qc/150201677.
[21] Moffat J W, "Misaligned Spin Merging Black Holes in Modified Gravity (MOG)", *Preprint* gr-qc/170605035.
[22] Moffat J W and Rahvar S, MNRAS **436**, 1439 (2013), *Preprint* astro-ph.GA/13066383.
[23] Moffat J W and Rahvar S, MNRAS **441**, 3724 (2014), *Preprint* astro-ph.CO/13095077.
[24] Moffat J W and Toth V T, Phys. Rev. **D91**, 043004 (2015), *Preprint* astro-ph.GA/14116701.
[25] Moffat J W, *Preprint* astro-ph.CO/151007037.
[26] Moffat J W, *Preprint* gr-qc/14102464.
[27] Shojai F, Cheraghchi S and Bouzari H Nezhad, Phys. Lett. **B 770**, 43 (2017), *Preprint* gr-qc/170404161.
[28] Cho H T, Cornell A S, Doukas J, Huang T R and Naylor W, Adv. Math. Phys. **2012**, 281705 (2012)
[29] Barakat T, Int. J. Mod. Phys. **A21**, 4127-4135 (2006).
[30] Zerilli F J, Phys. Rev. **D9**, 860-868 (1974).
[31] Moncrief V, Phys. Rev. **D9**, 2707-2709 (1974).
[32] Moncrief V, Phys. Rev. **D12**, 1526-1537 (1975).

IOP Conf. Series: Journal of Physics: Conf. Series **942** (2017) 012015 doi:10.1088/1742-6596/942/1/012015

# Finite conformal quantum gravity and spacetime singularities

**Leonardo Modesto and Lesław Rachwał**

Department of Physics, Southern University of Science and Technology, Shenzhen 518055, P. R. China

E-mail: lmodesto@sustc.edu.cn
E-mail: grzerach@gmail.com

**Abstract.** We show that a class of finite quantum non-local gravitational theories is conformally invariant at classical as well as at quantum level. This is actually a range of conformal anomaly-free theories in the spontaneously broken phase of the Weyl symmetry. At classical level we show how the Weyl conformal invariance is able to tame all the spacetime singularities that plague not only Einstein gravity, but also local and weakly non-local higher derivative theories. The latter statement is proved by a singularity theorem that applies to a large class of weakly non-local theories. Therefore, we are entitled to look for a solution of the spacetime singularity puzzle in a missed symmetry of nature, namely the Weyl conformal symmetry. Following the seminal paper by Narlikar and Kembhavi, we provide an explicit construction of singularity-free black hole exact solutions in a class of conformally invariant theories.

## 1. Introduction

There is one old symmetry that has always played a very crucial role in fundamental physics. This is the conformal symmetry, which is enjoyed by physical systems whose description does not depend on any characteristic scale. Historically this symmetry dates back to Hermann Weyl who discovered the concept of local gauge symmetries. At first, the gauge symmetries were applied to the local change of scale of the length measuring rulers. Since the measurements of space (and time) distances are the operational basis of general relativity (GR), we can understand why the conformal symmetry as a gauge symmetry first appeared in the gravitational setup. Only later the ideas of gauge symmetries were used in the non-gravitational setting of matter field theories, namely in the particle physics and condensed matter. Moreover, the conformal symmetry was found back in gauge theories too but the necessary condition here was to be in four spacetime dimensions. Indeed, the Maxwell's and Yang-Mills theories revealed to be exactly conformally invariant at the classical level. In the classical notion of conformal invariance the spacetime metric undergoes a rescaling by some always positive scalar function eventually dependent on the spacetime points. This is also the meaning of conformal symmetry used in classical GR, where it has been very useful to study the black holes' near-horizon geometry, the FRW cosmologies, and it was also a crucial element for the construction of the Newman-Penrose formalism.

There is also a different meaning of the same symmetry having to do with some properties of correlation functions on a given spacetime background. First of all, the change of scale on flat spacetime can be very easily transmitted to a change of coordinates and then the symmetry

IOP Conf. Series: Journal of Physics: Conf. Series **942** (2017) 012015 doi:10.1088/1742-6596/942/1/012015

shows up in a somehow global disguise. One can then study its extension to the full conformal group, which on a flat 4-dimensional Minkowski spacetime acts as a realization of the SO(2,3) group. This global symmetry initially finds its manifestation in some conformally invariant classical field theory models, but its true power is in constraining the correlation functions at the quantum level. Very soon, it was proven that the requirement of scale invariance at the quantum level implies that the beta functions of the couplings must exactly vanish. Moreover, the scale-invariance must be promoted to conformal symmetry to have a fully consistent field theory. The correlators of such very special theory show amazing symmetries that are described by the conformal algebra and the theories themselves are called conformal field theories (CFT's). When the beta functions are zero, then also the UV-divergences of the theory are absent, so the theory in question is UV-finite. There are very special examples of such theories and one of the most (super-) symmetric example is $\mathcal{N}=4$ Super-Yang-Mills theory in four dimensions. This line of development found its culmination in the AdS/CFT duality, where the ideas of CFT were also used to study the gravitational physics on the anti-de Sitter background and vice versa.

A natural question is whether these two ways of using conformal symmetry may be joined together. Since the conformal transformations are of spacetime character (not in the internal group of matter symmetries) then the merging should necessarily involve gravity. At the beginning, the crucial issue was how to make finite quantum gravity. Unfortunately, the quantum gravitational theories were from the beginning beset by UV-divergences and the first theory of gravity invented by Einstein in 1915 revealed to be non-renormalizable at the quantum level. There was completely no control over divergences, and so the theory had to be modified. The addition of terms with higher derivatives in the action helps with the renormalizability and even one-loop super-renormalizability was achieved [1]-[7]. In this way a much bigger control over divergences was gained. Additionally the inclusion of non-locality solved the unitarity issue. The last piece in this quest for UV-completion required the construction of a theory where all divergences were precisely cancelled. This was achieved for the first time by the authors in [8, 9, 10]. This result opened the possibility to have conformally invariant theories also at quantum level because the condition for the cancellation of the conformal anomaly is the vanishing of beta functions for all couplings. Now, we also see that the implementation of conformal symmetry in a fully gravitational and fully quantum framework has very profound and unexpected consequences. Indeed, not only we have a theory free of pathologies in the UV, but also we have exact spacetime solutions that are without singularities. This was another problem that hampered Einstein's GR, but now due to conformal quantum gravity we can firmly predict and describe what is beyond the "hypothetical" singularity at the black hole origin, Big Bang event, or in many others situations, which in the past created problems with predictability and extendability of the classical Cauchy evolution. The ideas of using conformal symmetry for solving these problems originated from the seminal work by Narlikar and Kembhavi [11].

Below we describe in some details the explicit construction of the finite quantum gravity with its natural conformal extension, and the mechanism for the resolution of spacetime singularities.

## 2. UV-finite quantum gravity

As motivated in the introduction Einstein's gravity must be modified to achieve UV-finiteness. We here consider a general gravitational theory described by the following Lagrangian density

$$\mathcal{L}_{\rm g} = -2\kappa_4^{-2}\sqrt{|g|}\left(R + R\gamma_0(\Box)R + R_{\mu\nu}\gamma_2(\Box)R^{\mu\nu} + R_{\mu\nu\rho\sigma}\gamma_4(\Box)R^{\mu\nu\rho\sigma} + \mathcal{V}(\mathcal{R})\right), \tag{1}$$

where the first two form-factors are defined in terms of entire functions $H_0$, $H_2$ by

$$\gamma_0(\Box) = -\frac{2\left(e^{H_0}-1\right)+4\left(e^{H_2}-1\right)}{12\Box} + \gamma_4(\Box) \quad \text{and} \quad \gamma_2(\Box) = \frac{e^{H_2}-1}{\Box} - 4\gamma_4(\Box)\,. \tag{2}$$

We remark that the third form-factor $\gamma_4(\Box)$ remains arbitrary and is constrained only by renormalizability properties of the theory. The functions $H_0$, $H_2$ are required to be analytic entire functions on the whole complex plane without poles and with asymptotic polynomial behaviour. The $\gamma_4$ is an analytic function without poles and with asymptotic behaviour compatible with the renormalizability of the theory. In $\mathcal{V}(\mathcal{R})$ we collect terms, which are cubic or higher powers in generalized gravitational curvatures $\mathcal{R}$. These terms are crucial in making the theory finite as we will show below.

The theory described by the Lagrangian (1) has the same symmetry as Einstein's gravity (diffeomorphism invariance). However, it is non-local due to the non-polynomial, but still analytic, structure in the form factors being functions of the covariant box operator [13, 14]. The two entire functions $H_0$ and $H_2$ fully specify the structure of the propagator for the gravitational fluctuations. If they satisfied the conditions stipulated above there are only two propagating degrees of freedom, namely the same as the two polarizations of massless gravitons in quantum Einstein's gravity. The theory is also unitary at perturbative level in Minkowski space.

For questions about renormalizability and UV divergences we need to study the behaviour of the theory for large values of momenta. We decided to concentrate on the theory, which in the UV regime tends to a polynomial higher derivative theory. This is achieved by requiring that for large values of the argument around real axis the form-factors $\gamma_i(\Box)$ for $i = 0, 2$ or $4$ have monomial asymptotics. The first examples of such form-factors were given in the seminal papers by Kuz'min and Tomboulis [2, 3, 12]. For simplicity we assume that the exponent there is a natural number $\gamma$ the same for all form-factors. It was found that for $\gamma \geqslant 3$ the theory is one-loop superrenormalizable [5, 6, 7], which means that only one-loop divergences survive and the theory from two loops on is completely convergent.

Having a super-renormalizable theory it is easy to look for an extension, in which all one-loop divergences are absent. For this purpose we use the terms in $\mathcal{V}(\mathcal{R})$. One can convince himself, using for example the covariant Barvinsky-Vilkovisky (BV) formalism for the quantum effective action, that the divergences can be cancelled by a precise choice of the front coefficients for the following operators,

$$s_1 R^2 \Box^{\gamma-2} R^2, \quad s_2 R^2_{\mu\nu} \Box^{\gamma-2} R^2_{\rho\sigma}, \quad s_3 R^2_{\mu\nu\rho\sigma} \Box^{\gamma-2} R^2_{\kappa\lambda\tau\omega} \tag{3}$$

Indeed, in four dimensions (and for asymptotically monomial form-factors) we have to make zero three beta functions for the dimensionless couplings in front of the operators quadratic in the gravitational curvatures, namely

$$\beta_{R^2}, \quad \beta_{R^2_{\mu\nu}}, \quad \beta_{R^2_{\mu\nu\rho\sigma}} \tag{4}$$

Actually due to a topological relation, using Gauss-Bonnet density, these beta functions are not independent (see also [15]). However, for full conformal invariance we have to secure that there are no divergent terms also at the boundary of our spacetime. It turns out that we have to cancel the fourth beta function

$$\beta_{\Box R} \quad \text{by using operators of the type} \quad s_4 (\Box R) \Box^{\gamma-2} R^2. \tag{5}$$

By using dimensional arguments, counting powers of covariant curvatures, and the power of BV formalism we see that the contribution of the listed above terms (4), (5) in $\mathcal{V}(\mathcal{R})$ to the corresponding beta function is linear in their coefficients $s_i$ $(i = 1, 2, 3, 4)$. The matrix giving such linear mapping can be easily shown to be non-degenerate hence the system of the equations for cancellation of these beta functions has always real solutions for the coefficients $s_i$. This construction bring us a UV-finite quantum gravitational theory.

IOP Conf. Series: Journal of Physics: Conf. Series **942** (2017) 012015 doi:10.1088/1742-6596/942/1/012015

## 3. Finite conformal quantum gravity

When we have a theory that behaves well at quantum level (namely does not have any divergences) we may compute its predictions to whatever accuracy with special care of finite terms. But computing these observables we notice that the theory possesses some hidden symmetry, which is revealed for example in the form of scattering amplitudes for on-shell gravitons [16]. This hidden symmetry is the Weyl's conformal invariance, and the main effort now is only to make such symmetry explicit in the action and on every step of the computation. In other words, we want to rewrite our finite quantum gravity in a conformally invariant way. We here use the compensating field method, which can be employed in the Lagrangian density (1) making the replacement $g^{\mu\nu} = \left(\phi^2\kappa_4^2\right)^{-1}\hat{g}^{\mu\nu}$, namely

$$\mathcal{L}_{\rm conf.} = -2\sqrt{|\hat{g}|}\left[\phi^2 R(\hat{g}) + 6\hat{g}^{\mu\nu}(\partial_\mu\phi)(\partial_\nu\phi)\right] \tag{6}$$
$$-\frac{2}{\kappa_4^2}\sqrt{|g|}\left[R(g)\,\gamma_0(\Box)R(g) + R_{\mu\nu}(g)\,\gamma_2(\Box)R^{\mu\nu}(g) + R_{\mu\nu\rho\sigma}(g)\,\gamma_4(\Box)R^{\mu\nu\rho\sigma}(g) + \mathcal{V}(\mathcal{R}(g))\right]\Big|_{g=\phi\hat{g}},$$

where by $|_{\phi\hat{g}}$ we mean that the metric $g^{\mu\nu}$ must be replaced with a product $\left(\phi^2\kappa_4^2\right)^{-1}\hat{g}^{\mu\nu}$.

In (6) we have an extra ghost-like degree of freedom. However, it can be eliminated using the extra symmetry of the theory, namely conformal invariance. Indeed, in this theory we have a dynamical metric tensor $\hat{g}_{\mu\nu}$ and a scalar field $\phi$. When the unitary gauge $\phi = $ const. is enplaned then we can come back to our finite quantum theory. Moreover, the quantum properties of the theory can not change in different gauges, and the theory must be UV-finite also when the scalar field is not completely gauged away, but another gauge is implemented.

We notice that the metric $g^{\mu\nu} = \left(\phi^2\kappa_4^2\right)^{-1}\hat{g}^{\mu\nu}$ (and therefore the action (6)) is invariant under the conformal rescalings

$$\hat{g}_{\mu\nu} \to \Omega^2(x)\,\hat{g}_{\mu\nu}\,, \quad \phi \to \Omega^{-1}(x)\,\phi, \tag{7}$$

meaning that $\phi$ describes clocks and rulers (physical scales) while $\hat{g}_{\mu\nu}$ contains exclusively information about the light-cone causal structure of the spacetime. After gauge fixing, $\phi$ can be just a constant and the distances are now measured by the metric $\hat{g}_{\mu\nu}$ or equivalently by the metric $g_{\mu\nu}$. The two metrics are now identified and, therefore, they have the same number of degrees of freedom, namely 10. By choosing a conformal gauge we select a vacuum for the field $\phi$ hence this process can be viewed as a spontaneous symmetry breaking. In other words the "effective number of d.o.f." is actually 10 before and after breaking of conformal invariance. The conformal "Higgs" mechanism here consists of moving the d.o.f., measuring time and space, from $\phi$ to the metric $\hat{g}_{\mu\nu}$. Such degree of freedom is the analog of the Higgs field giving mass to particles in the standard model of particle physics (SM).

The theory defined by the Lagrangian (6) can be analyzed like any other QFT. One can study the UV divergences of the theory when both the scalar and tensor perturbations propagate in any conformal gauge. We consider the fluctuations of the metric $\hat{g}_{\mu\nu}$ and of the scalar field $\phi$. Afterwords, we expand the action around a given spacetime background and a given background for the scalar field. The strict power counting and one-loop super-renormalizability of this theory can be easily shown, for more details we refer to [17]. Moreover, we have proved that the theory is unitary and all divergences are cancelled due to terms in $\mathcal{V}(\mathcal{R})$ regardless of the chosen conformal gauge in the quantization process. We also remind that the Fujikawa integration measure in path integrals is naturally conformally invariant. In the dimensional regularization scheme (DIMREG) some special care must be exerted when the theory is defined away from $d = 4$ spacetime dimensions. It is a known fact that the only way how conformal anomaly can arise is due to non-vanishing beta functions. If we do not have divergences, then we are also safe against conformal anomaly. This claim is independent whether the vacuum is in the broken or unbroken phase of the symmetry.

IOP Conf. Series: Journal of Physics: Conf. Series **942** (2017) 012015 doi:10.1088/1742-6596/942/1/012015

We showed that the theory is UV-finite (no divergences) and free of conformal anomaly (all the beta functions vanish), therefore, conformal invariance is preserved at the quantum level. The classical action of the theory (6) is also manifestly invariant, hence we have found a class of conformally invariant finite quantum gravity theories. These theories are in the spontaneously broken phase of the quantum conformal Weyl symmetry.

## 4. Resolution of the black hole singularity in conformal gravity

We propose a solution of the black hole singularity problem based on conformal symmetry. This idea was already explored in the 70's by Narlikar and Kembhavi [11], but we here reconsider and generalize their ideas for two reasons. First, we can use conformal methods because we have conformal quantum gravity, while they used it only at the classical level for a conformally invariant theory. Secondly, we made this idea to work not only for Big Bang singularities of FRW spacetimes (when the use of conformal methods seems rather trivial, cf. [19]), but foremost for black holes. Moreover, we can also get rid off of other pathologies like closed time-like curves, which by conformal rescalings can be reduced to have zero time lapse. Therefore, we conclude that we can make any singularity appearing in exact solutions of GR innocuous by a proper conformal rescaling.

Spacetime singularities are properties of some exact solutions of Einstein's gravity, where the group of symmetry is the group of diffeomorphisms. We remark that, contrary to some previous claims, theories with weak non-locality are also not able to get rid out of the spacetime singularities [18, 20, 21]. Weak non-locality is not enough and we need to enlarge the underlying diffeomorphism (Diff.) invariance of general relativity to include the Weyl conformal invariance. Indeed, only an enhanced symmetry, here local conformal symmetry is uniquely able to tame the singularities that plague not only Einstein gravity, but also local and weakly non-local higher derivative theories. We show this explicitly by analyzing the exact equations of motion (EOM).

We start with a Riemannian spacetime manifold $\mathcal{M}$ with a metric tensor $\hat{g}_{\mu\nu}$ and a scalar field $\phi$ as discussed in the previous section. Then the EOM for the metric $\hat{g}_{\mu\nu}$ in the theory (6) with the choice $\gamma_4 = 0$ read:

$$\phi^2 \hat{G}_{\mu\nu} = \nabla_\nu \partial_\mu \phi^2 - \hat{g}_{\mu\nu} \hat{\Box} \phi^2 - 6 \left( (\partial_\mu \phi)(\partial_\nu \phi) - \frac{1}{2} \hat{g}_{\mu\nu} g^{\alpha\beta} (\partial_\alpha \phi)(\partial_\beta \phi) \right)$$
$$- \frac{\delta \left( \sqrt{|g|} \left( R \gamma_0(\Box) R + R_{\mu\nu} \gamma_2(\Box) R^{\mu\nu} + \mathcal{V} \right) \right) \Big|_{\phi \hat{g}}}{\sqrt{|g_{\phi\hat{g}}|}\, \delta \hat{g}^{\mu\nu}}, \tag{8}$$

while the EOM for $\phi$ is:

$$\hat{\Box} \phi = \frac{1}{6} \hat{R} \phi - \frac{1}{\sqrt{|g_{\phi\hat{g}}|}} \frac{\delta \left( \sqrt{|g|} \left( R \gamma_0(\Box) R + R_{\mu\nu} \gamma_2(\Box) R^{\mu\nu} + \mathcal{V} \right) \right) \Big|_{\phi \hat{g}}}{\delta \phi}. \tag{9}$$

We can choose the operators in $\mathcal{V}(\mathcal{R})$ to be function of Ricci tensors only and not of the Riemann tensor. Since then all the operators resulting from the variation are at least linear in the Ricci tensor $R_{\mu\nu}$ (note that we here mean the Ricci tensor without a hat), then the Schwarzschild metric is an exact solution of the conformally invariant theory (6). This can be easily verified replacing the solution $\hat{g}_{\mu\nu} = g^{\rm Sch}_{\mu\nu}$ and $\phi = \kappa_4^{-1}$ in the above EOM (8) and (9). We notice as the result of the theorem that if the theory is general higher derivative or non-local then the same applies and we find Schwarzschild (with its singularity at the origin) as the exact solution of the theory [18].

However, the EOM are conformally invariant, hence if we consider another manifold $\mathcal{M}^*$ obtained from $\mathcal{M}$ by a conformal transformation (7) to $^*$ quantities, then also $\hat{g}^*_{\mu\nu}$ and $\phi^*$ satisfy the EOM (8) and (9). The transformation $\phi \to \phi^*$ is mathematically valid provided $\Omega^{-1}$ does not vanish (or become infinite). It is assumed that $\Omega = \Omega(x)$ is a twice differentiable positive function (or a smooth function in a non-local theory) of the spacetime coordinates $x$.

Starting from the exact Schwarzschild solution we can now construct an infinite class of other exact solutions, all of them conformally equivalent to the Schwarzschild metric, by doing rescalings (7) to both the scalar field $\phi$ and the metric $\hat{g}_{\mu\nu}$ with some $\Omega$. For $\phi = \kappa_4^{-1}$ and $\Omega = 1$ we get the Schwarzschild spacetime Moreover, $\hat{R}_{\mu\nu}(\hat{g}^*_{\mu\nu}) \neq 0$, so the conformally rescaled Schwarzschild metric is not any more Ricci-flat.

In this large class of solutions we can pick out a subclass of singularity-free black hole spacetimes. One example of which is provided by the following singularity-free exact black hole solution (in any conformally invariant theory), obtained by rescaling the Schwarzschild metric by a suitable $\Omega$:

$$ds^{*2} = \left(1 + \frac{L^4}{r^4}\right)\left[\left(1 - \frac{2m}{r}\right)dt^2 + \frac{dr^2}{1 - \frac{2m}{r}} + r^2 d\Omega^{(2)}\right], \quad \phi^* = \left(1 + \frac{L^4}{r^4}\right)^{-1/2} \kappa_4^{-1}. \qquad (10)$$

If this metric is considered in GR, then one can easily show that the spacetime is regular at $r = 0$. This is supported by the regularity of all curvature invariants (like Kretschmann scalar) and by the geodesic completion for massive as well as massless particles moving in this geometry [22]. In the solution (10) $L$ is an arbitrary length scale. This can be naturally chosen either to be the Planck length or the mass of the black hole $m$. The warp factor $\Omega(x)$ chosen in (10) is one of the infinitely many possible which render the transformed spacetime completely regular and without any singularities. An analogous rescalings can be found to make regular Kerr-Newman rotating and charged black hole backgrounds. In a conformally invariant theory this metric turns out to be physical and observable if the conformal symmetry is spontaneously broken.

In conformal gravity the black hole thermodynamics and the Hawking evaporation process had been presented in [23]. Similarly, the issue with the black hole entanglement entropy in the quantum field theory framework had been solved in [24]. Basically, the black hole entropy is finite because the theory (1) is UV-finite at the quantum level.

## References

[1] Asorey M, López J L , Shapiro I L, *Int. J. Mod. Phys.* A 12 5711 (1997) (*Preprint* hep-th/9610006)
[2] Tomboulis E T, (*Preprint* hep-th/9702146)
[3] Tomboulis E T, *Mod. Phys. Lett.* A **30**, no. 03n04, 1540005 (2015)
[4] Khoury J, *Phys. Rev.* D **76**, 123513 (2007) (*Preprint* hep-th/0612052)
[5] Modesto L, *Phys. Rev.* D **86**, 044005 (2012) (*Preprint* hep-th/1107.2403)
[6] Modesto L, *Astron. Rev.* 8.2 (2013) 4-33 (*Preprint* hep-th/1202.3151)
[7] Modesto L, *Preprint* hep-th/1202.0008
[8] Modesto L and Rachwal L, *Nucl. Phys.* B **889**, 228 (2014) (*Preprint* hep-th/1407.8036)
[9] Modesto L and Rachwal L, *Nucl. Phys.* B **900**, 147 (2015) (*Preprint* hep-th/1503.00261)
[10] Modesto L, Piva M and Rachwal L, *Phys. Rev.* D **94**, no. 2, 025021 (2016) (*Preprint* hep-th/1506.06227)
[11] Narlikar J V and Kembhavi A K, *Lett. Nuovo Cim.* **19**, 517 (1977)
[12] Kuz'min Y V, *Yad. Fiz.* **50**, 1630 (1989) [*Sov. J. Nucl. Phys.* **50**, 1011 (1989)].
[13] Modesto L and Rachwal L, *Int. J. Mod. Phys.* D **26**, 1730020 (2017)
[14] Calcagni G and Nardelli G, *Phys. Rev.* D **82**, 123518 (2010) (*Preprint* hep-th/1004.5144)
[15] Modesto L, Rachwal L and Shapiro I L, *Preprint* hep-th/1704.03988
[16] Dona P, Giaccari S, Modesto L, Rachwal L and Zhu Y, *J. High Energy Phys.* JHEP08(2015)038
[17] Modesto L and Rachwal L, *Preprint* hep-th/1605.04173
[18] Modesto L and Rachwal L, *J. High Energy Phys.* JHEP1512(2015)173 (*Preprint* hep-th/1506.08619)
[19] Koshelev A S, Modesto L, Rachwal L and Starobinsky A A, *J. High Energy Phys.* JHEP1611(2016)067
[20] Calcagni G and Modesto L, *Phys. Lett.* B **773**, 596 (2017) (*Preprint* gr-qc/1707.01119)
[21] Modesto L, Moffat J W and Nicolini P, *Phys. Lett.* B **695**, 397 (2011) (*Preprint* gr-qc/1010.0680)
[22] Bambi C, Modesto L and Rachwal L, *J. Cosmol. Astropart. Phys.* JCAP **1705**, no. 05, 003 (2017)
[23] Bambi C, Modesto L, Porey S and Rachwal L, *J. Cosmol. Astropart. Phys.* JCAP **1709**, 033 (2017)
[24] Giaccari S, Modesto L, Rachwal L and Zhu Y, *Preprint* hep-th/1512.06206

IOP Conf. Series: Journal of Physics: Conf. Series **942** (2017) 012016 doi:10.1088/1742-6596/942/1/012016

# Thermodynamical aspects of black holes in modified gravity

**Lorenzo Sebastiani**

TIFPA - INFN, Via Sommarive 14, 38123 Povo (TN), Italy
Dipartimento di Fisica, Università di Trento,Via Sommarive 14, 38123 Povo (TN), Italy

E-mail: lorenzo.sebastiani@unitn.it

**Abstract.** We discuss some thermodynamical definitions for black holes in modified theories of gravity.

## 1. Introduction

In General Relativity (GR), several thermodynamical notions can be introduced for the black holes (BHs), but in the modified theories of gravity the black hole solutions are not expected to share the same proprieties of their Einsteinian counterparts. In $F(R)$-modified gravity the First law of thermodynamics can be derived from the equations of motion, evaluating independently the entropy via Wald method and the Killing-Hawking temperature from the metric, and an expression for the BH Killing energy can be found. In an analogue way, in other theories of modified gravity (for instance, in Gauss-Bonnet gravity) the First Law of thermodynamics can be used to infer the black hole energy.

This proceeding is mainly based on Refs. [1, 2, 3].

## 2. Black holes in General Relativity

Any spherically symmetric and four dimensional metric can be locally expressed in the form:

$$ds^2 = \gamma_{ij}(x^i)dx^i dx^j + \mathcal{R}^2(x^i)d\Omega_2^2\,, \qquad i,j \in \{0,1\}\,, \tag{1}$$

where $d\Omega_2^2$ is the metric of a two-dimensional maximally symmetric space, $\gamma_{ij}(x^i)$ is the reduced metric of the normal space-time with coordinates $x^i$ and $\mathcal{R}(x^i)$ is the areal radius and is a function of the coordinates of the normal space. On the normal space one can introduce the scalar quantity

$$\chi(x^i) = \gamma^{ij}(x^i)\partial_i\mathcal{R}(x^i)\partial_j\mathcal{R}(x^i)\,, \tag{2}$$

such that the sphere with areal radius $\mathcal{R}(x^i)$ turns out to be trapped when $\chi(x^i) < 0$; marginal when $\chi(x^i) = 0$; untrapped when $0 < \chi(x^i)$. Thus, the dynamical trapping horizon of a black hole is defined by the conditions

$$\chi(x^i)\Big|_{\rm H} = 0\,, \quad 0 < \partial_i\chi(x^i)|_{\rm H}\,. \tag{3}$$

In this paper, the pedex 'H' denotes a quantity evaluated on the coordinates of the horizon.

In General Relativity we can associate to the black hole horizons several thermodynamical quantities, namely the energy, the entropy and the surface gravity. For the energy we have a quasi-local definition given by the Misner-Sharp formula,

$$E_{MS}(x^i) := \frac{1}{2G_{\rm N}}\mathcal{R}(x^i)\left[1-\chi(x^i)\right] , \tag{4}$$

with $G_{\rm N}$ the Newton's constant. Thus, the Misner-Sharp energy evaluated on the BH horizon corresponds to the BH Killing energy/mass,

$$E = \frac{r_{\rm H}}{2G_{\rm N}} . \tag{5}$$

The entropy of a black hole satisfies the Area Law,

$$S = \frac{\mathcal{A}_{\rm H}}{(4G_{\rm N})} , \tag{6}$$

namely is proportional to the area $\mathcal{A}_{\rm H}$ of the horizon.

Finally, for static black holes we may use the time-like Killing vector field $\xi_\mu(x^\nu)$ to define the Killing surface gravity $\kappa_K$ as follows,

$$\kappa_K \xi^\mu(x^\nu) = \xi^\nu \nabla_\nu \xi^\mu(x^\nu) . \tag{7}$$

In the dynamical case, where the time-like Killing vector field is absent, Hayward found a way to infer the surface gravity by working with the metric only,

$$\kappa_K := \frac{1}{2}\Box_\gamma \mathcal{R}(x^i)\Big|_{\rm H} , \tag{8}$$

where the d'Alambertian is evaluated with respect to the reduced metric $\gamma_{ij}(x^i)$.

It is possible to associate a temperature to the horizon of a black hole. In fact the black holes are not so black and may emit radiation, dubbed the "Hawking radiation", due to the quantum effects near to the horizon [4]. In the static case, all derivations of the Hawking radiation rate leads to the semi-classical expression,

$$\Gamma \equiv {\rm e}^{-\frac{2\pi \Delta E_K}{\kappa_K}} , \tag{9}$$

in terms of the change $\Delta E_K$ of the Killing energy of the emitted particle and the Killing surface gravity [5]. Thus, the surface gravity can be identified with the Killing temperature as

$$T_K := \frac{\kappa_K}{2\pi} . \tag{10}$$

Therefore, if one uses the change of the entropy $\Delta S$ one easily obtains the First Law of black hole thermodynamics,

$$\Delta E_K = T_K \Delta S . \tag{11}$$

In the dynamical case Hayward found a way to derive the First Law from the equations of motion. Assuming the Einstein's equation of GR, in a generic four-dimensional spherically symmetric space-time, the following geometric identity holds true on the black hole trapping horizon [6],

$$\frac{\kappa_K}{2\pi}\frac{d}{dr_{\rm H}}\left(\frac{\mathcal{A}_{\rm H}}{4G_{\rm N}}\right) = \frac{d}{dr_{\rm H}}\left(\frac{\mathcal{R}_{\rm H}}{2G_{\rm N}}\right) + \frac{{\rm T}^{(2)}_{\rm H}}{2}\frac{d}{dr_{\rm H}}\mathcal{V}_{\rm H} , \tag{12}$$

IOP Conf. Series: Journal of Physics: Conf. Series **942** (2017) 012016 doi:10.1088/1742-6596/942/1/012016

where $\mathcal{V}_{\rm H}$ is the three-volume enclosed by the horizon and $\mathrm{T}^{(2)}_{\rm H}$ is the reduced trace of the matter stress-energy tensor at the horizon and acts like a working term. On thermal equilibrium the Gibbs equation leads to,

$$T\Delta S = \Delta E + pdV\,, \quad dV = \mathcal{V}_k r_{\rm H}^2 dr_{\rm H}\,, \tag{13}$$

such that, by introducing the entropy (6) and the BH energy (5), one may suggest the Kodama/Hayward temperature,

$$T_H := \frac{\kappa_{\rm H}}{2\pi}\,. \tag{14}$$

Let us restrict our analysis to the Static Spherically Symmetric (SSS) space-time. The metric reads,

$$ds^2 = -\mathrm{e}^{2\alpha(r)}B(r)dt^2 + \frac{dr^2}{B(r)} + r^2 d\Omega_k^2\,, \quad d\Omega_k^2 = \left(\frac{d\rho^2}{1-k\rho^2} + \rho^2 d\phi^2\right)\,, \tag{15}$$

where $\alpha(r)$ and $B(r)$ are functions of the radial coordinate only, $\mathcal{R} = r$ is the areal radius and the topology depends on the $k$ parameter and can be spherical, flat or hyperbolic for $k = +1, 0, -1$, respectively. A static solution describes a black hole as soon as there exists an event horizon with a real and positive radius $r = r_{\rm H}$ where

$$B(r_{\rm H}) = 0\,, \quad 0 < B'(r)|_{r_{\rm H}}\,. \tag{16}$$

The prime index corresponds to the derivative with respect to $r$. We should note that in the static case the Killing temperature $T_K$ and the Kodama temperature $T_H$ associated to the event horizon are in principle different when $\alpha(r_H) \neq 0$,

$$T_K := \frac{1}{4\pi}\mathrm{e}^{\alpha(r_{\rm H})}B'(r_{\rm H})\,, \quad T_H := \frac{1}{4\pi}B'(r_{\rm H})\,. \tag{17}$$

In General Relativity this is not a problem. The Hawking radiation rate is independent on the choices of temperature and energy of the emitted particle and in the vacuum case of the Schwarzshild solution one has $\alpha(r) = 0$ and the two definitions coincide. However, this is not true for the vacuum case of a modified gravity theory where $\alpha(r) \neq 0$. Moreover, in modified gravity is not easy to define the energy of a black hole. In General Relativity the Misner-Sharp mass corresponds to the charge of a conserved current from the second-order differential equations of the theory, but in modified gravity we deal with higher derivative field equations and we must use a different approach. In what follows, we will consider some classes of modified theories with black hole solutions and we will analyze the First Law of Thermodynamics in their framework.

## 3. $F(R)$-four dimensional modified gravity

Let us consider the $F(R)$-gravity in vacuum, whose action is given by (see Ref. [7] for some general reviews),

$$I = \frac{1}{16\pi G_{\rm N}}\int_{\mathcal{M}} d^4x\sqrt{-g}F(R)\,. \tag{18}$$

Here, $g$ is the determinant of the metric tensor $g_{\mu\nu}(x^\mu)$, $F(R)$ is a function of the Ricci scalar only and $\mathcal{M}$ is the space-time manifold. Given a static black hole solution described by the metric (15), if and only if $R_{\rm H}$ explicitly depends on $r_{\rm H}$ only, from the (0,0)-component of the $F(R)$-field equations evaluated on the event horizon we obtain,

$$T_k\Delta S_W = \mathrm{e}^{\alpha(r_{\rm H})}\mathcal{V}_k\left(\frac{k\,F_R(R_{\rm H})}{2G_{\rm N}} - \frac{R_{\rm H}F_R(R_{\rm H}) - F(R_{\rm H})}{4G_{\rm N}}r_{\rm H}^2\right)\,, \tag{19}$$

where $\mathcal{V}_k \equiv \mathcal{A}_k/r^2$ depends on the topology, the Killing temperature $T_K$ (17) emerges in a natural way, and $S_W$ is the Wald entropy [8],

$$S_W = \frac{\mathcal{A}_k(r_{\rm H})\,F_R(R_{\rm H})}{4G_{\rm N}}\,, \quad \Delta S_W = \frac{1}{4G_{\rm N}}\left(2V_k r_{\rm H} F_R(R_{\rm H}) dr_{\rm H} + V_k r_{\rm H}^2 F_{RR}(R_{\rm H}) dR_{\rm H}\right) . \tag{20}$$

The second expression above holds true when $R_{\rm H}$ is an explicit function of $r_{\rm H}$ only. The condition on the entropy looks quite restrictive, but in a large class of explicit examples of $F(R)$-static black hole solutions it is well satisfied. Thus, we can derive for a generic $F(R)$-gravitational model a First Law of black hole thermodynamics in the form

$$\Delta E_K := T_K \Delta S_W\,, \tag{21}$$

where $\Delta E_K$ is the variation of the Killing energy of the black hole itself. As a consequence, one may define

$$E_K := \frac{\mathcal{V}_k}{4\pi}\int e^{\alpha(r_{\rm H})}\left(\frac{k\,F_R(R_{\rm H})}{2G_{\rm N}} - \frac{R_{\rm H}F_R(R_{\rm H}) - F(R_{\rm H})}{4G_{\rm N}} r_{\rm H}^2\right) dr_{\rm H}\,. \tag{22}$$

Here, an expression for the BH-energy is proposed by deriving the First Law from the equations of motion of $F(R)$-gravity, evaluating independently the entropy via Wald method and the Hawking temperature via quantum mechanical methods in curved space-times.

Let us consider some examples where only one integration constant $C$ appears in the SSS metric which may describe a black hole for some choices of the topology. For the case $R = 4\Lambda$ with $\Lambda = (RF_R(R) - F)/(2F_R(R))$ and Schwarzshild dS/AdS solution $\alpha(r) = 0$, $B(r) = (k - C/r - \Lambda r^2/3)$, we get

$$T_K = \frac{4\pi r_{\rm H}}{(1-\Lambda r_{\rm H}^2)}\,, \quad S_W = \frac{\mathcal{A}_k(r_{\rm H})F_R(R_{\rm H})}{4G_{\rm N}}\,, \quad E_K = \frac{\mathcal{V}_k F_R(R_{\rm H})}{8\pi G_{\rm N}} r_{\rm H}\left(k - \frac{\Lambda}{3}r_{\rm H}^2\right) . \tag{23}$$

Therefore, by using the fact that $B(r_{\rm H}) = 0$, one has

$$E_K = \frac{\mathcal{V}_k F_R(R_{\rm H})}{8\pi G_{\rm N}}\, C\,. \tag{24}$$

For the model $F(R) = \gamma\sqrt{k(R+12\lambda)}$ with $\alpha(r) = 0$ and $B(r) = \left(\frac{k}{2} - \frac{C}{r^2} + \lambda r^2\right)$, the BH Killing energy reads $E_K \propto C$.

For the model $F(R) = \gamma(1/R - h^2/6)$ with $\exp[2\alpha] = r/r_0$, $r_0$ being a dimensional parameter, and $B(r) = \frac{4}{7}\left(k - \frac{7r}{6h} + \frac{C}{r^{7/2}}\right)$, one obtains $E_K \propto C$.

For the class of Clifton-Barrow models $F(R) = R^{\delta+1}(\kappa^2)^\delta$, $\delta \neq 1$, the metric reads [9],

$$ds^2 = -\left(\frac{r}{r_0}\right)^{2a}\left(k - \frac{C}{r^b}\right)dt^2 + \frac{\beta dr^2}{\left(k - \frac{C}{r^b}\right)} + r^2 d\Omega_k^2\,, \tag{25}$$

where $a, b$ and $\beta$ are functions of $\delta$. Also in this case the BH Killing energy results to be $E_K \propto C$. In all these examples, the Killing energy is proportional to the integration constant of the metric, giving to it a physical meaning like in the Schwarzwschild case of GR. We point out that, when $\alpha(r) \neq 0$, if one uses the Hayward prescription such a result cannot be achieved, but with the Killing formalism there are some sorts of cancellations and we obtain these reasonable results.

Let us consider the case of $R^2$-gravity, where two integration constants appear in the metric. The action reads:

$$I = \int_{\mathcal{M}} d^4x\sqrt{-g}\,R^2\,. \tag{26}$$

IOP Conf. Series: Journal of Physics: Conf. Series **942** (2017) 012016 doi:10.1088/1742-6596/942/1/012016

Such a model is often considered in the inflationary scenario and admits the Schwarzshild dS/AdS solution,

$$\alpha(r) = 0\,, \quad B(r) = \left(k - \frac{C}{r} - \frac{\lambda r^2}{3}\right)\,, \tag{27}$$

where $R = 4\lambda$ and the cosmological constant $\lambda$ is a free parameter like $C$, due to the fact that it is not fixed by the gravitational Lagrangian. As a consequence, when we take the thermodynamical variation of the Killing energy of the black hole described by this solution, we must consider the variation with respect to $\lambda$ also and it is not possible to give an explicit expression for the energy. On the other hand, the Wald entropy of the black hole reads,

$$S_W = \frac{\mathcal{A}_k(r_\mathrm{H})R_\mathrm{H}}{2} = 2\mathcal{A}_k(r_\mathrm{H})\lambda\,, \tag{28}$$

and vanishes for $\lambda = 0$. The cosmological constant plays the role of the inverse of the Planck Mass of GR, since the action is scale invariant. In fact, $\lambda = 1/L^2$ introduces a fundamental lenght scale into the theory and one may consider it like a fixed parameter. Only in this case the First Law leads to

$$E_K = \frac{\mathcal{V}_k}{\pi}\lambda\left[kr_\mathrm{H} - \frac{\lambda r_\mathrm{H}^3}{3}\right] = \frac{\mathcal{V}_k}{\pi}\lambda C\,. \tag{29}$$

We observe that the presence of the $R^2$-term modifies the energy of a Schwarzschild dS/AdS black hole when the cosmological constant is different to zero (for example, in the model with Lagrangian $\mathcal{L} = [(R - 2\Lambda)/(16\pi G_N) + R^2]$).

## 4. Gauss-Bonnet modified gravity

Let us consider now the following action,

$$I = \frac{1}{16\pi G_\mathrm{N}}\int_\mathcal{M} d^4\sqrt{-g}F(R,G)\,, \tag{30}$$

where $F(R,G)$ is a function of the Ricci scalar $R$ and the Gauss-Bonnet four dimensional topological invariant $G$. In this framework some static SSS black hole solutions are known, but in general it is not possible to derive the First Law from the field equations of the theory. However, given a BH solution, it is still possible to evaluate its Killing temperature, its Wald entropy and therefore its Killing energy. For example, the model with $F(R,G) = R + \sqrt{G}$ admits the topological SSS solution,

$$ds^2 = -B(r)dt^2 + \frac{dr^2}{B(r)} + r^2 d\left(\frac{d\rho^2}{1 - k\rho^2} + \rho^2 d\phi^2\right)\,, \quad B(r) = -k + \frac{r}{C}\,, \tag{31}$$

$C$ being an integration constant, which describes a black hole in the spherical topological case with $k = 1$. The Wald entropy reads

$$S_W = \frac{\mathcal{A}_1(r_\mathrm{H})}{4G_\mathrm{N}}\left[\mathcal{F}'_R + \mathcal{F}'_G\left(\frac{4k}{r^2}\right)\right]\Big|_\mathrm{H} = \frac{\pi r_\mathrm{H}^2}{G_\mathrm{N}}\left[1 + \frac{C}{r_\mathrm{H}}\right]\,, \tag{32}$$

and we see that, since $\partial_r S_W|_{r_\mathrm{H}} \neq \Delta S$, the First Law can not been derived from the equations of motion like in $F(R)$-gravity. On the other side, by using the First Law with the Killing temperature we find

$$T_K = \frac{1}{4\pi C}\,, \quad E_K = \frac{C}{G_\mathrm{N}}\,, \tag{33}$$

and we see that even in this case the integration constant of the solution can be identified with the energy. We may conclude that in the vacuum case of $F(R)$ and $F(R,G)$-gravity the Killing formalism leads to resonable definitions for the thermodynamics of the black holes.

IOP Conf. Series: Journal of Physics: Conf. Series **942** (2017) 012016 doi:10.1088/1742-6596/942/1/012016

## 5. Non vacuum static spherically symmetric solution

As a last example, we will consider the following model where a scalar field $\phi$ is non-minimally coupled with the electromagnetic potential ($F_{\mu\nu}$ is the electromagnetic field strength tensor):

$$I = \int_{\mathcal{M}} d^4x\sqrt{-g}\left[\frac{(R-2\Lambda)}{16\pi G_N} - \frac{1}{2}\partial^\mu\phi\partial_\mu\phi + V(\phi) - \xi e^{\sqrt{16\pi G_n}\lambda\phi}(F^{\mu\nu}F_{\mu\nu})\right]\,, \quad V(\phi) = V_0 e^{\gamma\sqrt{16\pi G_N}\phi}\,. \tag{34}$$

Here, $\Lambda\,,\lambda,\ \xi\,,\gamma$ and $V_0$ are fixed parameters of the theory. This model admits the following class of topological Lifshitz-like solutions,

$$ds^2 = -\left(\frac{r}{r_0}\right)^z B(r)dt^2 + \frac{dr^2}{B(r)} + r^2 d\Omega_k^2\,, \tag{35}$$

where $z$ is a number and $r_0$ a dimensional parameter. The equations of motion constrain the field $\phi = \phi(r)$ as

$$\phi(r) = \sqrt{\frac{2z}{16\pi G_N}}\log[r/r_0]\,. \tag{36}$$

The form of $B(r)$ results to be,

$$B(r) = \frac{2k}{z+2} - \frac{C}{r^{1+\frac{z}{2}}} + \frac{\tilde{V}_0 r^2}{(2\gamma\sqrt{2z}+6+z)}\left(\frac{r_0}{r}\right)^{\lambda\sqrt{2z}} + \frac{8\xi\tilde{Q}^2}{(2\lambda\sqrt{2z}+2-z)r^2}\left(\frac{r}{r_0}\right)^{\gamma\sqrt{2z}} - \frac{4\Lambda r^2}{6+z}\,, \tag{37}$$

where $C$ is a free integration constant of the solution, $\tilde{V}_0 = 16\pi G_N V_0$, $\tilde{Q}^2 = G_N Q^2$, $Q$ being the charge of the electromagnetic potential, and the parameters of the model must be related to each other in order to satisfy the Klein-Gordon equation of the scalar field. For $\phi(r) = 0$, namely $z = 0$, $\xi = 1/4$ and $V_0 = 0$, we recover the Reissner-Norstrom solution with cosmological constant.

The model under investigation has second order field equations like in GR and when solution (35, 37) describes a black hole its mass is well defined as

$$E = \frac{\mathcal{V}_k}{8\pi G_N}C\,, \tag{38}$$

while the entropy satisfies the Area Law (6). Now from the first component of the field equations evaluated on the BH horizon we derive

$$\frac{1}{4\pi}B'(r)|_H\Delta S = \Delta E + pdV\,, \quad p = (p_\phi + p_{\rm EM})_{\rm radial}\,,$$

where the working term collects the contributions from the radial pressures of the scalar and electromagnetic fields. Thus, the First Law holds true by making use of the Hayward temperature in (17). It looks that when we consider a static but non vacuum solution the Hayward formalis is more consistent to describe the thermodynamics of the black holes.

## References

[1] L. Sebastiani and S. Zerbini, Eur. Phys. J. C **71**, 1591 (2011).
[2] G. Cognola, O. Gorbunova, L. Sebastiani and S. Zerbini, Phys. Rev. D **84**, 023515 (2011).
[3] R. Myrzakulov, L. Sebastiani and S. Zerbini, Gen. Rel. Grav. **45**, 675 (2013) [arXiv:1208.3392 [gr-qc]].
[4] S. W. Hawking, Nature **248** 30 (1974); Commun. Math. Phys. **43** 199-220 (1975).
[5] M. K. Parikh and F. Wilczek, Phys. Rev. Lett. **85** (2000) 5042 [hep-th/9907001].
[6] S. A. Hayward, Class. Quant. Grav. **15**, 3147 (1998) [gr-qc/9710089]; S. A. Hayward, R. Di Criscienzo, L. Vanzo, M. Nadalini and S. Zerbini, Class. Quant. Grav. **26**, 062001 (2009) [arXiv:0806.0014 [gr-qc]].
[7] S. Nojiri and S. D. Odintsov, eConf C **0602061** (2006) 06 [Int. J. Geom. Meth. Mod. Phys. **4** (2007) 115] [hep-th/0601213]; S. Nojiri and S. D. Odintsov, Phys. Rept. **505** (2011) 59 [arXiv:1011.0544 [gr-qc]].
[8] R. M. Wald, Phys. Rev. D **48**, no. 8, R3427 (1993) doi:10.1103/PhysRevD.48.R3427 [gr-qc/9307038].
[9] T. Clifton and J. D. Barrow, Phys. Rev. D **72**, no. 10, 103005 (2005) [gr-qc/0509059].

IOP Conf. Series: Journal of Physics: Conf. Series **942** (2017) 012017 doi:10.1088/1742-6596/942/1/012017

# Testing general relativity with X-ray reflection spectroscopy of MCG-06-30-15

**Ashutosh Tripathi**

Center for Field Theory and Particle Physics and Department of Physics, Fudan University, 200438 Shanghai, China

E-mail: ashutosh_tripathi@fudan.edu

**Abstract.** The spacetime geometry around astrophysical black holes is well approximated by the Kerr metric, but deviations from standard predictions are possible in a number of scenarios beyond Einstein's gravity and in the presence of exotic matter. In this paper, we present the constraints on possible deviations from the Kerr solution using X-ray reflection spectroscopy from the analysis of real data. We use the relativistic X-ray reflection code RELXILL modified to a generic stationary, axisymmetric and asymptotically flat black hole metric. We analyze 350 ks long XMM-Newton observations of the AGN in MCG-06-30-15 taken during July-August 2001 and constrain the Johannsen deformation parameter $\alpha_{13}$.

## 1. Introduction

The general theory of relativity was proposed by Albert Einstein in 1915 and it has been used to describe the dynamics of spacetime since then. Several experiments, like binary pulsar timing array, calculation of light deviation by Sun etc., has been performed in order to test the general relativity in weak gravity regime [1]. The theory is standard in the sense that the theoretical predictions from the theory agree well with experimental deductions. In order to test deviation from general relativity in describing the dynamics of spacetime, the theory needs to be tested in strong gravity regime. In other words, one need to design an experiment to test this theory where gravity is very strong. The ideal probe in the universe to test general relativity in strong gravity regime are black holes [2].

In Einstein gravity, an uncharged black hole is described by Kerr solution and, only two parameters, namely, mass (M) and angular momentum (J), are required for its complete quantification [3]. Although Kerr metric should describe a black hole, deviations from it are also possible.

There are two distinct methods through which general relativity can be tested; Gravitational and electromagnetic radiation [4, 5, 6]. Electromagnetic approach asssumes the geodesic motion of Kerr metric for analysis and can be easily generalized to test the deviation [4, 5]. This approach analyze the properties of electromagnetic radiation from the accretion disk close to black hole. It depends on the motion of gas in strong gravity region and photon propagation from the emission point in the disk to the faraway region where spacetime is almost flat.

There are two leading electromagnetic techniques for testing the Kerr metric hypothesis; continuum fitting method and the reflection method [2]. Reflection method involves the analysis of relativistically smeared reflection spectrum of thin accretion disk. It can be applied to both stellar mass black holes and supermassive black holes. It is independent of black hole mass and distance. The inclination angle of the disk w.r.t. to line of sight is calculated by fitting the reflection spectrum. For continuum method, these three quantities must be inferred from other methods.

In this work, we employ a test metric, Johannsen metric [7], which is a generalized form of Kerr metric to describe spacetime around black hole. The deviations from Kerr metric are quantified by deformation parameters which are constrained using astrophysical data. Here, we use XMM-Newton 2001 observations of the supermassive black hole MCG-06-30-15.

## 2. The source MCG-06-30-15

MCG-06-30-15 is a narrow line seyfert 1 (NLSy1) galaxy situated at a distance of 37 Mpc ($z = 0.008$). This source has shown significant deviation from simple power-law based on the evidence from both absorption and emission feature from cold iron line [14]. The first X-ray observation of this source was reported with *ASCA*, which was taken during 1993 July-August [15]. The warm absorber edges and iron emission lines are clearly resolved, for the first time, using this observation. They, also, discovered large change in column density between two observations taken during this period. The detection of relativistic effects in X-ray emission line ($K_\alpha$ line) from ionized iron in this source was first reported in [16]. The observed $\mathrm{K}_\alpha$ line is broad and antisymmetric because of the relativistic effects, which are so strong that it was believed to be originated from the innermost region of accretion disk. Iwasawa et al. [17] analysed the variability of the source during *ASCA* observation. The peculiar line profile during deep minimum state suggests that the line emitting region is very close to a central spinning black hole where enormous gravitational effect operates.

Multi-wavelength study of this source was explained in [18], which confirms the presence of double zone warm absorber. The analysis at softer energies is presented in [19] for an energy range of 0.32-1.7 KeV and they found a reasonable fit using multizone dusty warm absorber.

Brenneman & Reynolds presented the analysis of iron line and warm absorber of 2001 XMM-Newton observation in [9]. Their new model, KERRDISK, allows black hole spin to be a free parameter and found the value of spin to be $0.989 \pm 0.002$. The spectral variability was examined in [20] using simultaneous XMM-Newton and NuSTAR observations taken in 2013. The extreme variability of the source can be explained in terms of intrinsic X-ray fluctuations and light bending model.

The relativistic broadening of iron line in this source make it suitable for studying the deviation from Kerr metric.

## 3. Observations and Data Reduction

MCG-06-30-15 was observed by XMM-Newton [21] on 31 July, 2 August and 4 August 2001 for a total period of 350 ks during three revolutions 301,302 and 303 respectively. The observations were taken with EPIC pn and MOS1/2 cameras operated in small window and medium filter mode [22]. Data from MOS detectors are not analyzed because they suffered significantly from pile-up.

We used *XMM-Newton* Science Analysis Software (SAS) version 16.0.0 for data reduction. Three EPIC pn fits files of different observation IDs were combined to form one single fits file using ftool *MERGE*. The good time intervals (GTIs) were generated using *TABTIGEN* which contains data free of proton flares. The processed event files were generated by combining GTI and event list files and, then, filtered using single and double events (PATTERN $\leq$ 4) but excluding events that are at the edge of CCD (FLAG=0). To obtain source event files, we have extracted an area of radius 40 arcsec. For background events, an area of radius 50 arcsec, located as far away as possible from the source, is chosen to avoid any of its contribution. After

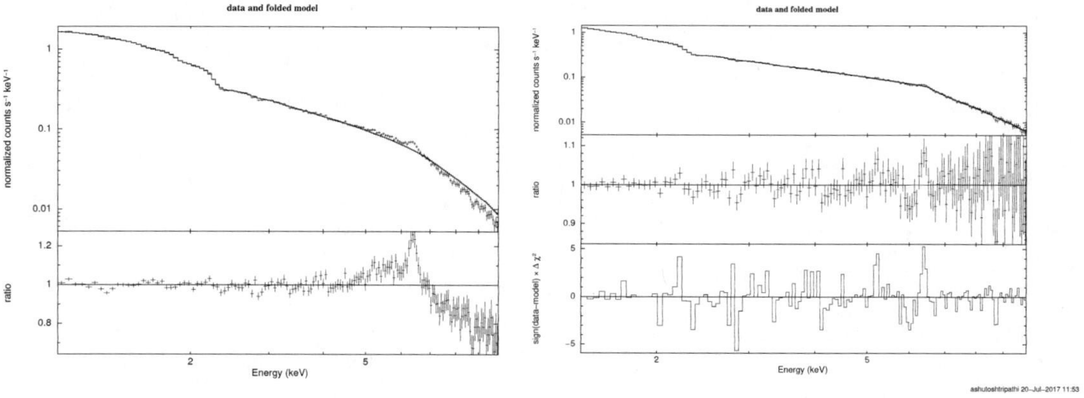


**Figure 1.** *Left panel* : The 2001 XMM-Newton observation ratioed against the power law. *Right Panel* : The data to model ratio for the model employed in this paper.

backscaling source and background spectra, response files and ancillary files were generated using rmfgen and arfgen, respectively. Lastly, In order to oversample instrumental resolution by a factor of 3 and to have minimum counts of 30 per background-subtracted spectral channel so that $\chi^2$ statistics can be used, the spectra were rebinned using *SPECGROUP*.

## 4. Theoretical Models

In figure 1, we show the ratio of data to power law continuum in order to identify different features in the data, which we have tried to analyze with our theoretical models in the present work. The source is highly variable throughout the entire energy range. Most prominent feature is the emission line at 6.4 KeV, which is likely due to iron $K_\alpha$ emission.
In order to model the variability in soft energy range ($<$ 2.0 KeV), we employ double zoned warm absorber. Physically, warm absorbers are ionized absorbing gas believed to be originated from accretion disk. It surrounds the central engine of AGN like a torus (see for e.g. [9] and references therein). They contain a rich forest of lines and edges from various species of gas and dust. They incorporate several "zones" of materials, distinct in their kinematic properties as well as their column densities and ionizations, but maintained in pressure balance. For computation, it is approximated as discrete zones characterized by a column density and ionization parameter. XSTAR spectral synthesis package for photoionized gases is used to construct a grid of warm absorbers as a function of column density and ionization parameter. They are multiplicative models, absorber models that can be applied to any emission spectrum.

The disk produces a whole spectrum of recombination and flourescent lines. To obtain realistic constraints, we need to model the entire X-ray reflection spectrum (see, e.g., [2] and references therein). For X-ray reflection, XILLVER provides the best treatment of radiative transfer using photoionization routines from XSTAR which contains the most detailed atomic database for modelling photoionized X-ray spectra. XILLVER provides a detailed treatment of K-shell atomic properties of ionized ions. The most prominent one is iron $K_\alpha$ line at 6.4 KeV. This line is narrower in emitter's reference but it becomes broad and skewed in receiver's frame of reference because of general relativistic effects (Doppler effect, gravitational redshift) which are prominent in strong gravity regime. These effects are so strong that the emission lines can possibly be radiated from innermost regions of black hole.

RELCONV is a relativistic convolution code assuming Kerr metric. It gives the spectrum

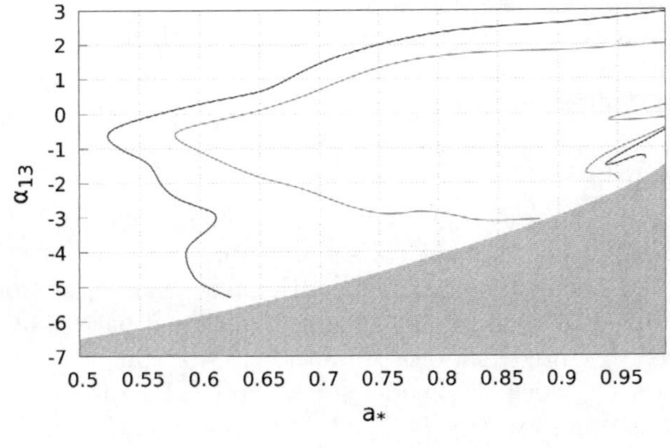


**Figure 2.** 90% and 95 % confidence contours in $a^* - \alpha_{13}$ plane.

| Components | Parameters | Description | Best fit value |
|---|---|---|---|
| TBABS | $N_H$ $(10^{22})$ | Column density for Galactic absorption | 0.29 ±0.06 |
| WARMABS1 | $N_H$ $(10^{22})$ | Column density for I warm absorber | 0.24 ±0.11 |
| | $log\xi$ | Ionization parameter for I warm absorber | 1.40 ±0.34 |
| WARMABS2 | $N_H$ $(10^{22})$ | Column density for II warm absorber | 0.40±0.27 |
| | $log\xi$ | Ionization parameter for II warm absorber | 2.47 ±0.27 |
| RELXILL_NK | $Index1$ | Emissivity index for $r < Rbr$ | 2.3±0.28 |
| | $Index2$ | Emissivity index for $r > Rbr$ | 3.0* |
| | $R_{br}$ | Break radius | 76.60 ± 121.18 |
| | $a$ | spin of Black hole | 0.92±0.12 |
| | $i(deg)$ | Inclination angle w.r.t. line of sight | 3.6 ±35.8 |
| | $\Gamma$ | Power law Index of the primary source ($E^{-\Gamma}$) | 1.98±0.03 |
| | $log(\xi_{refl})$ | Ionization parameter at the inner edge of disk | 3.34±0.04 |
| | $Afe$ | Iron abundance in solar units | 2.21* |
| | $refl\_frac$ | Reflection fraction | 1.72±0.57 |
| | $defpar\_value$ | Deformation parameter $\alpha_{13}$ | -0.82±2.5 |

**Table 1.** Best fit parameters of different components of the model used and their description. The parameters marked with * are frozen during analysis. The errors quoted are at 90% confidence interval.

measured by a distant observer given the local spectrum at any emission point in the disk. The code RELXILL is the combination of RELCONV and XILLVER. It describes relativistic smearing of X-ray reflection spectrum near the black hole [10, 11, 12]. RELXILL_NK is the extension of RELXILL to Johannsen metric, of which Kerr metric is a special case [13]. In order to test Kerr metric, RELXILL_NK must describe the X-ray reflection spectrum in a background metric, more general than Kerr metric, which contains some deformation parameters in addition to mass (M) and angular momentum (J).

For modelling spectrum and variability, following model is used to fit the observation in XSPEC

$$\text{Model 1}: \quad \text{TBABS} * \text{WARMABS1} * \text{WARMABS2} * \text{RELXILL\_NK}. \tag{1}$$

RELXILL_NK models power law continuum and reflection component from region near black hole and take into account the relativistic smearing of iron line. The cut-off energy of input spectrum is taken to be 300 KeV. The inner edge of the disk is assumed to be located at the radius of innermost stable circular orbit ($R_{isco}$). WARMABS1 and WARMABS2 are two warm absorbers used to model ionized gas around the source. TBABS models the galactic absorption.

## 5. Results and Discussion

In table 1, we report the best fit parameters for the model which is used to fit the spectrum in the energy range of 1.3-10.0 KeV. The reduced chi square value ($\chi^2_{red}$) is found to be 1.04. The value of spin is found to be 0.92 which is consistent with earlier claims of this source being a rapidly spinning black hole. Due to degeneracies of iron-line model parameters, some parameters could not be constrained. Figure 2 shows 90% and 95% confidence contours in spin- $\alpha_{13}$ plane. The Kerr metric is recovered for this model as the null value of deformation parameter is included in confidence ellipses. However, it could not constrain the spin effectively. This is because of high variability of the source and the presence of complex absorption and emission features, which could not be modelled properly with warm absorber and galactic absorption.

## 6. Conclusions and Future Work

We tested the deviations from general relativity using astrophysical data. This is achieved by modifying the existing X-ray reflection model, which assumes Kerr metric, to a more generic metric. We obtained constraints on deformation parameters by analyzing 350 ks long XMM-Newton observation of MCG-06-30-15 which is famous for its relativistically broadened iron line. Unfortunately, the current model cannot constrain the parameters properly because of highly variable nature of the source. Other sources can also be analyzed in order to have better constraints on deformation parameters [24]. Future X-ray missions like ATHENA and LAD/eXTP [23] can be proved useful in constraining possible deviations from Kerr metric.

## Acknowledgments

This research is based on observations obtained with *XMM-Newton*, an ESA science mission with instruments and contributions directly funded by ESA Member States and NASA.

AT acknowledges support from the China Scholarship Council (CSC), grant no. 2016GXZR89.

## References

[1] C. M. Will, Living Rev. Rel. **17**, 4 (2014) [arXiv:1403.7377 [gr-qc]].
[2] C. Bambi, *Black Holes: A Laboratory for Testing Strong Gravity* (Springer Singapore, 2017), doi:10.1007/978-981-10-4524-0
[3] P. T. Chrusciel, J. L. Costa and M. Heusler, Living Rev. Rel. **15**, 7 (2012) [arXiv:1205.6112 [gr-qc]].
[4] C. Bambi, Rev. Mod. Phys. **89**, 025001 (2017) [arXiv:1509.03884 [gr-qc]].
[5] C. Bambi, J. Jiang and J. F. Steiner, Class. Quant. Grav. **33**, 064001 (2016) [arXiv:1511.07587 [gr-qc]].
[6] N. Yunes and X. Siemens, Living Rev. Rel. **16**, 9 (2013) [arXiv:1304.3473 [gr-qc]].
[7] T. Johannsen, Phys. Rev. D **88**, no. 4, 044002 (2013) [arXiv:1501.02809 [gr-qc]].
[8] A. C. Fabian, K. Iwasawa, C. S. Reynolds and A. J. Young, Publ. Astron. Soc. Pac. **112**, 1145 (2000) [astro-ph/0004366].
[9] L. W. Brenneman and C. S. Reynolds, Astrophys. J. **652**, 1028 (2006) [astro-ph/0608502].
[10] T. Dauser, J. Garcia, J. Wilms, M. Bock, L. W. Brenneman, M. Falanga, K. Fukumura and C. S. Reynolds, Mon. Not. Roy. Astron. Soc. **430**, 1694 (2013) [arXiv:1301.4922 [astro-ph.HE]].
[11] J. Garcia, T. Dauser, C. S. Reynolds, T. R. Kallman, J. E. McClintock, J. Wilms and W. Eikmann, Astrophys. J. **768**, 146 (2013) [arXiv:1303.2112 [astro-ph.HE]].
[12] J. Garca *et al.*, Astrophys. J. **782**, 76 (2014) [arXiv:1312.3231 [astro-ph.HE]]
[13] C. Bambi, A. Cardenas-Avendano, T. Dauser, J. A. Garcia and S. Nampalliwar, Astrophys. J. **842**, 76 (2017) [arXiv:1607.00596 [gr-qc]].
[14] Nandra, K., Pounds, K. A., Stewart, G. C., Fabian, A. C., Rees, M. J. 1989. Mon. Not. Roy. Astron. Soc. **236**, 39P-46P.

IOP Conf. Series: Journal of Physics: Conf. Series **942** (2017) 012017 doi:10.1088/1742-6596/942/1/012017

[15] Fabian, A. C. *et al.* 1994. Publications of the Astronomical Society of Japan **46**, L59-L63.
[16] Tanaka, Y. *et al.* 1995. Nature **375**, 659-661.
[17] Iwasawa, K. *et al.* 1996. Mon. Not. Roy. Astron. Soc. **282**, 1038-1048.
[18] Reynolds, C. S., Ward, M. J., Fabian, A. C., Celotti, A. 1997. Mon. Not. Roy. Astron. Soc. **291**, 403.
[19] Turner, A. K., Fabian, A. C., Vaughan, S., Lee, J. C. 2003. Mon. Not. Roy. Astron. Soc. **346**, 833-840.
[20] Marinucci, A.*et al.* 2014. The Broadband Spectral Variability of MCG-6-30-15 Observed by NuSTAR and XMM-Newton. The Astrophysical Journal **787**, 83.
[21] Jansen, F. *et al.* 2001. Astronomy and Astrophysics **365**, L1-L6.
[22] Turner, M. J. L. *et al.* 2001. Astronomy and Astrophysics **365**, L27-L35.
[23] S. N. Zhang *et al.* [eXTP Collaboration], Proc. SPIE Int. Soc. Opt. Eng. **9905**, 99051Q (2016) [arXiv:1607.08823 [astro-ph.IM]].
[24] Z. Cao, S. Nampalliwar, C. Bambi, T. Dauser and J. A. Garcia, arXiv:1709.00219 [gr-qc].

IOP Conf. Series: Journal of Physics: Conf. Series **942** (2017) 012018 doi:10.1088/1742-6596/942/1/012018

# Shear viscosity in QFTs dual to AdS spherical black holes

**Matteo Tuveri, Mariano Cadoni and Edgardo Franzin**
Dipartimento di Fisica, Università di Cagliari & INFN, Sezione di Cagliari,
Cittadella Universitaria, 09042 Monserrato, Italy

E-mail: matteo.tuveri@ca.infn.it

**Abstract.** In the framework of the AdS/CFT correspondence, we define and compute the spherical analogue of the shear viscosity for QFTs dual to five-dimensional charged AdS black holes in general relativity (GR) and Gauss-Bonnet (GB) gravity. We show that the ratio between this quantity and the entropy density, $\tilde{\eta}/s$, exhibits a temperature-dependent hysteresis.

## 1. Introduction

Nowadays, holography plays a central role in theoretical gravitational physics and has produced several successful applications as, for example, the AdS/CFT correspondence. Moreover it gives insights on quantum gravity and it is a powerful tool for the description of phase transitions and the computation of transport coefficients in strongly coupled QFTs [1–5] in the hydrodynamic limit. In particular, it has been proposed that the shear viscosity to entropy density ratio $\eta/s$ satisfies a fundamental bound $\eta/s \geqslant 1/4\pi$, known as the KSS bound [1], which found support both from string theory [2] and quark-gluon plasma experimental data [6]. By now, it is well-known that the KSS bound is violated by higher curvature terms in the Einstein-Hilbert action [7] or by breaking of translational or rotational symmetry of the black brane background [8–12]. Typically, when the KSS bound is violated, $\eta/s$ exhibits a non-trivial dependence on the temperature [13].

Until now, these investigations have been restricted to planar topologies in the bulk (black branes) and have not concerned spherical topologies (black holes). The main obstruction to this generalisation is the absence of the usual hydrodynamic limit for QFTs dual to spherical black holes (BHs). Indeed, differently from the black brane case, the spherical geometry of the horizon breaks the translational symmetry in the dual QFT preventing the existence of conserved charges. However, it is still possible to define a relativistic hydrodynamics in curved spacetimes without translational symmetry as an expansion in the derivatives of the hydrodynamic fields of the stress-energy tensor [14] and a related Kubo formula for the shear viscosity.

In this work we discuss whether it is possible to use transport coefficients of the dual QFT to learn about the complicated thermodynamical phase portrait of BHs, in particular five-dimensional AdS BHs coupled to an electromagnetic field in GR [15] and GB gravity [16].

## 2. Hydrodynamics in curved spacetimes

The hydrodynamic limit of a QFT living on a curved spacetime can be defined in the same way as for a QFT in the plane. We just consider the system at large relaxation times (small

IOP Conf. Series: Journal of Physics: Conf. Series **942** (2017) 012018 doi:10.1088/1742-6596/942/1/012018

frequencies) and large scales compared to the microscopic scale of the system. When the latter is unknown, we can still give a thermal description of the system and associate this microscopic scale with the inverse of the temperature $T$. Thus, the hydrodynamic limit corresponds to consider excitations of the system with wavelength $\lambda \gg 1/T$. In this limit, the macroscopic behaviour of the QFT living in a curved background is described by a stress-energy tensor, which can be written as [14, 17]

$$T^{ab} = (\epsilon + P)\, u^a u^b + P g^{ab} + \Pi^{ab}\,, \tag{1}$$

where $\epsilon$ and $P$ are the energy density and the thermodynamical pressure and $u^a$ is the fluid velocity, usually considered in the frame in which the fluid is at rest. The tensor $\Pi^{ab}$ contains all the dissipative contributions to the stress-energy tensor. At first order in the velocity expansion, it depends on three transport coefficients $\kappa$, $\tau_\Pi$ and $\eta$, the latter known as the shear viscosity.

The previous considerations hold for a QFT in a generic curved space. Working in the AdS/CFT framework, we can apply Eq. (1) to a four dimensional CFT dual to a five-dimensional AdS bulk spherical BH [18–20]. To derive a Kubo formula for CFTs living on the boundary of $AdS_5$, whose spatial section is the three-sphere, we consider small perturbations around the boundary background metric, i.e. $g_{ab} = \bar{g}_{ab} + h_{ab}$. In general we can consider three different types of perturbations: shear, sound and transverse (scalar) modes. The behaviour of these modes will be encoded in three different correlators $G_{1,2,3}(\omega, k)$, where $k$ is the momentum. In the translationally invariant case (and also when translation invariance is broken by external matter fields) at $k = 0$ these three functions are equal, owing to rotational symmetry [7]. By contrast, in the spherical case under consideration, the momentum cannot be taken to zero by construction (see later for details) and the correlators will be different. Thus, in general, any definition of the shear viscosity in a spherical background based on linear response to small disturbances will be channel-dependent [21]. In this work we will focus on the transverse (and traceless) perturbations. The computations for the sound and shear channel is left for future investigations. Under these assumptions and in linear approximation, Eq. (1) becomes

$$T^{ij} = -P h_{ij} - \eta \dot{h}_{ij} + \eta \tau_\Pi \ddot{h}_{ij} - \frac{\kappa}{2}\left(\ddot{h}_{ij} + L^2 \Delta_L h_{ij}\right), \tag{2}$$

where $\Delta_L$ is the Lichnerowicz operator and $L$ is the $AdS_5$ length. We choose a harmonic time dependence for the perturbation and we expand it in hyperspherical harmonics. We now extract the retarded Green function for the spatial components of the stress-energy tensor $T^{ij}$ in the tensor channel and, from Eq. (2), we read

$$G^R_{T^{ij}T^{ij}}(\omega, \ell) = -P - i\omega\eta - \omega^2 \eta \tau_\Pi - \frac{\kappa}{2}\left(\omega^2 + L^2 \gamma\right), \tag{3}$$

where $\gamma \equiv \ell(\ell + 2) - 2$ is the eigenvalue of the Lichnerowicz operator and $\ell = 1, 2, 3, \ldots$ is the first number associated with the hyperspherical harmonic expansion. Equation (3) allows us to derive a Kubo formula for the analogue of the shear viscosity $\tilde{\eta}$ for a relativistic QFT on a spatial spherical background as

$$\tilde{\eta} = -\lim_{\omega \to 0} \frac{1}{\omega} \operatorname{Im} G^R_{T^{ij}T^{ij}}(\omega, \ell \to \ell_0)\,, \tag{4}$$

where $\ell_0$ is the smallest eigenvalue of the Lichnerowicz operator and $\omega$ is the frequency of the perturbation. Notice that the only difference of Eq. (4) with the planar case is the evaluation of the retarded Green function in $\ell \to \ell_0$ instead of wavenumber $k \to 0$.

It is important to stress that, with respect to the planar case, we have an additional contribution to the stress-energy tensor (2) ruled by the transport coefficient $\kappa$. This is rather expected in view of the breaking of translational invariance. However, these additional contributions do not contribute to the shear viscosity.

## 3. Black hole solutions in five dimensions

We consider static and spherically symmetric five-dimensional BHs,

$$ds^2 = g^{(0)}_{ab}\, dx^a dx^b = -f(r)\, dt^2 + \frac{dr^2}{f(r)} + r^2\, d\bar{\Omega}_3^2\,, \tag{5}$$

where $d\bar{\Omega}_3^2$ is the line element of the 3-sphere. For the AdS-RN BHs the metric function is

$$f_{\rm RN}(r) = 1 + \frac{r^2}{L^2} - \frac{8G_5 M}{3\pi r^2} + \frac{4\pi G_5 Q^2}{3r^4}\,, \tag{6}$$

while, in the branch that allows for BH solutions, the metric function for GB gravity is [22–24]

$$f_{\rm GB}(r) = 1 + \frac{r^2}{2\lambda L^2}\left[1 - \sqrt{1 - 4\lambda L^2\left(\frac{1}{L^2} - \frac{8G_5 M}{3\pi r^4} + \frac{4\pi G_5 Q^2}{3r^6}\right)}\right]. \tag{7}$$

In Eqs. (6) and (7), $L$ is the AdS length, $M$ and $Q$ are, respectively, the BH mass and charge. To have asymptotically AdS-GB BH solutions, the dimensionless GB coupling constant $\lambda$ must be smaller than $1/4$. Its value is also constrained by the unitarity bounds for the dual QFT [7,25,26], so that in this work we consider the range $0 < \lambda \leqslant 9/100$.

The temperature of these BHs can be expressed in terms of the horizon radius $r_+$

$$T(r_+) = \left(\frac{r_+}{\pi L^2} + \frac{1}{2\pi r_+} - \frac{2G_5 Q^2}{3r_+^5}\right)\left(1 + \frac{2\lambda L^2}{r_+^2}\right)^{-1}. \tag{8}$$

For RN BHs, as the charge of the BH decreases to the critical charge $Q_c = L^2/6\sqrt{5\pi}$, the BH undergoes a second-order phase transition. Below $Q_c$ the system is characterised by the presence of two stable states (small and large BHs) connected through a meta-stable region of intermediate BHs. The phase transition small/large BHs is a first-order one [27]. The picture is analogous for GB BHs [28,29]. This is very similar to a Van der Waals-like fluid behaviour with second-order phase transition controlled by a critical parameter ($\lambda$ and/or $Q$ for GB BHs, $Q$ for RN BHs) [27,30], first-order one (controlled by $T$) and metastabilities (small/large BH region separated by metastable region of intermediate BHs).

## 4. Linear perturbations

Following the rules of the AdS/CFT correspondence, to compute the spherical analogue viscosity to entropy density ratio $\tilde{\eta}/s$ for the QFT dual to a five-dimensional spherically symmetric charged AdS BH, we consider transverse and traceless perturbations about the background metric (5), $g_{ab} = g^{(0)}_{ab} + h_{ab}$. In particular, $h_{ab} = 0$ unless $(a,b) = (i,j)$ and $h_{ij}(r,t,x) = r^2\,\phi(r,t)\,\bar{h}_{ij}(x)$ being $\bar{h}_{ij}$ the eigentensor of the Lichnerowicz operator built on the background 3-sphere, whose eigenvalues are $\ell(\ell+2) - 2$ with $\ell = 1, \ldots$. Such perturbations are gauge-invariant and by linearising the Einstein field equations, the angular part decouples [31, 32]. By assuming a harmonic time dependence for the perturbation, $\phi(r,t) = \psi(r)\, e^{-i\omega t}$, one finds the linear second-order differential equation for $\psi(r)$

$$\frac{1}{r^3}\frac{d}{dr}\left[r^3 f(r) F(r)\frac{d\psi(r)}{dr}\right] + \left[\frac{F(r)}{f(r)}\omega^2 - m^2(r)\right]\psi(r) = 0\,, \tag{9}$$

where $f(r)$ is given by Eq. (6) for AdS-RN BHs or Eq. (7) for GB BHs, $F(r) \equiv 1 - \lambda L^2 f'(r)/r$. The mass term for the perturbation is $m^2(r) = (4 - \ell(\ell+2))\,(1 - \lambda L^2 f''(r))/r^2$ and $\ell$ are the

IOP Conf. Series: Journal of Physics: Conf. Series **942** (2017) 012018 doi:10.1088/1742-6596/942/1/012018

eigenvalues of the Lichnerowicz operator on the 3-sphere. The presence of a non-vanishing mass term in Eq. (9) is a consequence of the breaking of translational symmetry due to the spherical geometry of the horizon. The general solution of Eq. (9) does not exist in analytical form, but in the $r \to \infty$ limit, Eq. (9) with $\omega = 0$ admits as solutions a non-normalisable and a normalisable mode that behave asymptotically as

$$\psi_0(r) = 1 - \frac{\lambda L^2}{2\left(1-\sqrt{1-4\lambda}\right) r^2} + \mathcal{O}\left(\log r/r^4\right), \quad \psi_1(r) = \frac{1}{r^4} + \mathcal{O}\left(1/r^6\right). \tag{10}$$

The near-horizon behaviour of the non-normalisable mode is different for non-extremal and extremal BHs. For the former case, it is given by a power-series expansion, i.e. $\psi_0(r) = \psi_0(r_+) + \mathcal{O}(r - r_+)$, while in the latter $\psi_0(r) = (r - r_+)^\nu$ being $\nu$ an appropriate index.

## 5. The shear viscosity to entropy density ratio

The retarded Green function in Eq. (4) can be found using a modified version of the method proposed in Refs. [9,33] to include higher-order curvature corrections. The method gives a very simple and elegant way for computing correlators in a QFT dual to a gravitational bulk theory. The analogue shear viscosity to density entropy ratio is determined by the non-normalisable mode $\psi_0(r)$ evaluated at the horizon times a function of $\lambda$

$$\frac{\tilde{\eta}}{s} = \frac{1}{4\pi}\psi_0(r_+)^2 \left[1 - 4\lambda\left(1 - \frac{2\pi G_5 Q^2 L^2}{3r_+^6}\right)\right]\left(1 + \frac{6\lambda L^2}{r_+^2}\right)^{-1}. \tag{11}$$

In the large temperature regime, by inverting Eq. (8) and using Eq. (10), Eq. (11) becomes

$$\frac{\tilde{\eta}}{s} = \frac{1-4\lambda}{4\pi}\left[1 - \frac{\lambda L^2\left(7 - 6\sqrt{1-4\lambda}\right)}{\pi^2\left(1-\sqrt{1-4\lambda}\right) L^4 T^2} + \mathcal{O}\left(1/T^4\right)\right]. \tag{12}$$

In the $T \to \infty$ limit, the value of $\tilde{\eta}/s$ tends to $(1-4\lambda)/4\pi$ which reduces to the KSS bound for $\lambda = 0$. These bounds are in general violated as Eq. (12) is a decreasing function of the temperature. In the extremal case, the metric function and its first derivative vanish when evaluated on the horizon as well as the non-normalisable mode $\psi_0(r_+)$. This means that $\tilde{\eta}/s$ goes to zero in the $T = 0$ extremal limit as $\tilde{\eta}/s \sim T^{2\nu}$.

The global behaviour of $\tilde{\eta}/s$ as a function of $T$ is obtained as follows. For each value of the charge and the GB parameter, there exists a minimum mass— and hence a minimum radius. Then we numerically integrate Eq. (9) with $\omega = 0$ outwards from the horizon to infinity supplied with a power-series boundary condition for $\psi_0(r)$. Next, we use a shooting method to determine $\psi_0(r_+)$ by requiring that $\psi_0(\infty) = 1$. Finally, the temperature and $\tilde{\eta}/s$ for each solution are computed with Eqs. (8) and (11). Units $G_5 = L = 1$ are adopted.

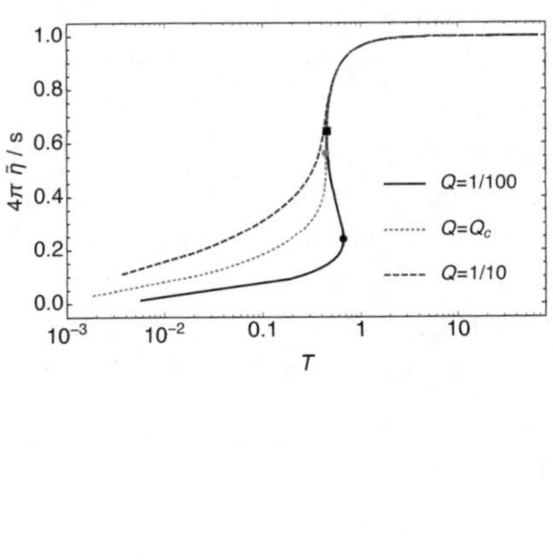


**Figure 1.** Behaviour of $\tilde{\eta}/s$ as a function of the temperature for dual QFTs of AdS-RN BHs. We plot $\tilde{\eta}/s$ for selected values of the BH charge above, at and below the critical value. The dots and squares mark the critical temperatures relative to the small/large BH first-order phase transition. We have considered the smallest eigenvalue of the Lichnerowicz operator, i.e. $\ell = \ell_0 = 1$.

IOP Conf. Series: Journal of Physics: Conf. Series **942** (2017) 012018 doi:10.1088/1742-6596/942/1/012018

In Fig. 1 we plot our results for $\tilde{\eta}/s$ as a function of the temperature for QFTs dual to AdS-RN BHs for selected values of the charge $Q$ and we observe that, for $Q < Q_c$, it exhibits a temperature-dependent hysteresis, after that the second-order Van der Waals-like phase transition occurs. In Fig. 2 we plot our numerical results for $\tilde{\eta}/s$ as a function of $T$ for QFTs dual to AdS-GB BHs. We show three different cases: neutral BHs; fixed $\lambda$; fixed charge $Q$.

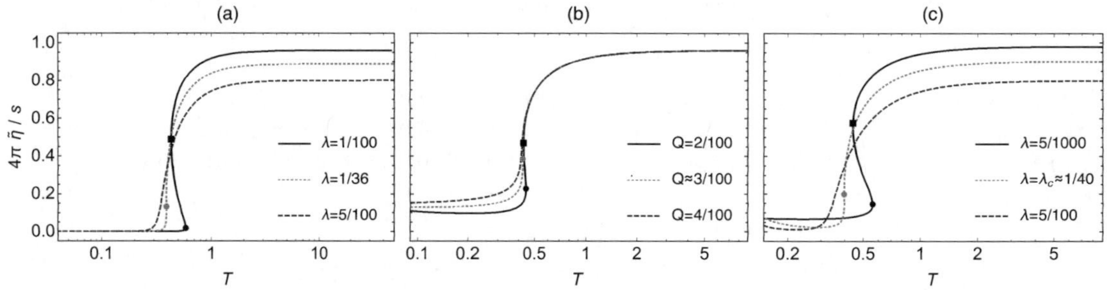


**Figure 2.** Behaviour of $\tilde{\eta}/s$ as a function of the temperature for GB BHs with (a) $Q = 0$ for selected values of the GB coupling constant above, at and below the critical value; (b) $\lambda = 1/100$, and selected values of charge above, at and below the critical value; (c) $Q = 1/100$, and selected values of GB constant above, at and below the critical value. Dots (squares) mark the maximum (minimum) of the temperature as a function of the BH horizon.

In the hydrodynamic, holographic context, a hysteretic behaviour in the shear viscosity has been already observed for AdS black branes with broken rotational symmetry and with a p-wave holographic superfluid dual [8]. Moreover, it is known that real fluids may exhibit hysteresis in the $\eta$-$T$ plane, this is, for instance, the case of nanofluids [34].

The mechanism that generates hysteresis in $\tilde{\eta}/s$ is the same that is responsible for the phase transition and can be traced back to non-equilibrium thermodynamics. An unstable (metastable) region of intermediate BHs connects two stable regions of large and small BHs. A potential barrier prevents the evolution of the system from occurring as an equilibrium path between the two stable states [35, 36]. Equilibrium will be reached passing through a metastable region and a path-dependence of $\tilde{\eta}/s$ is generated.

Notice that when the breaking of translational symmetry is generated by external fields, the symmetry may be restored or not when the system flows to the IR (as in the black brane case) [9]. In the BH case instead, because the breaking has a geometric and topological origin, translational symmetry cannot be restored in the IR.

## 6. Conclusion

We have used the AdS/CFT correspondence to obtain information about the behaviour of bulk BHs by studying the hydrodynamic properties of the dual QFTs. In particular, we have focussed on the scalar channel and defined and computed the shear viscosity to entropy ratio for QFTs holographically dual to five-dimensional AdS BH solutions of GR and GB gravity.

Our most important result is the behavior of $\tilde{\eta}/s$ at intermediate temperatures: for AdS-RN BHs a second-order Van der Waals phase transition occurs as the system goes from large values of $Q$ to the critical one; below the critical charge the BH undergoes a temperature-dependent first-order phase transition and $\tilde{\eta}/s$ develops hysteresis. A similar behaviour occurs for GB gravity when we fix the charge and let the GB coupling constant $\lambda$ to vary and viceversa.

Our definition of $\tilde{\eta}/s$ is channel dependent. In general we have three different determinations of $\tilde{\eta}$ for shear, sound and transverse (scalar) perturbations. In this work we have focused on the transverse ones. However our results suggest that, for QFTs dual to bulk spherical BHs, the hydrodynamical, long wavelength modes can be described by the $\ell \to \ell_0$ modes that probe large angles on the sphere (cfr. $k \to 0$ on the plane). Due to the spherical topology, the hydrodynamic interpretation in terms of conserved quantities fails. However, we can still define the shear viscosity through a Kubo formula where the stress-energy tensor is only covariantly conserved and interpret it as the rate of entropy production due to a strain.

In holographic models, the shear viscosity to entropy ratio of the QFT is closely related and keeps detailed information about the thermodynamical phase structure of the dual BH background.

## References

[1] Kovtun P, Son D T and Starinets A O 2003 *J. High Energy Phys.* JHEP10(2003)064 (*Preprint* hep-th/0309213)
[2] Kovtun P, Son D T and Starinets A O 2005 *Phys. Rev. Lett.* **94** 111601 (*Preprint* hep-th/0405231)
[3] Policastro G, Son D T and Starinets A O 2002 *J. High Energy Phys.* JHEP09(2002)043 (*Preprint* hep-th/0205052)
[4] Hartnoll S A, Herzog C P and Horowitz G T 2008 *Phys. Rev. Lett.* **101** 031601 (*Preprint* 0803.3295)
[5] Horowitz G T 2011 *Lect. Notes Phys.* **828** 313 (*Preprint* 1002.1722)
[6] Song H, Bass S A, Heinz U, Hirano T and Shen C 2011 *Phys. Rev. Lett.* **106** 192301 (*Preprint* 1011.2783)
[7] Brigante M, Liu H, Myers R C, Shenker S and Yaida S 2008 *Phys. Rev.* D **77** 126006 (*Preprint* 0712.0805)
[8] Erdmenger J, Kerner P and Zeller H 2011 *Phys. Lett.* B **699** 301 (*Preprint* 1011.5912)
[9] Hartnoll S A, Ramirez D M and Santos J E 2016 *J. High Energy Phys.* JHEP03(2016)170 (*Preprint* 1601.02757)
[10] Alberte L, Baggioli M and Pujolas O 2016 *J. High Energy Phys.* JHEP07(2016)074 (*Preprint* 1601.03384)
[11] Burikham P and Poovuttikul N 2016 *Phys. Rev.* D **94** 106001 (*Preprint* 1601.04624)
[12] Cadoni M, Frassino A M and Tuveri M 2016 *J. High Energy Phys.* JHEP05(2016)101 (*Preprint* 1602.05593)
[13] Cremonini S, Gürsoy U and Szepietowski P 2012 *J. High Energy Phys.* JHEP08(2012)167 (*Preprint* 1206.3581)
[14] Baier R, Romatschke P, Son D T, Starinets A O and Stephanov M A 2008 *J. High Energy Phys.* JHEP04(2008)100 (*Preprint* 0712.2451)
[15] Cadoni M, Franzin E and Tuveri M 2017 *Phys. Lett.* B **768** 393 (*Preprint* 1702.08341)
[16] Cadoni M, Franzin E and Tuveri M 2017 *Preprint* 1703.05162
[17] Romatschke P 2010 *Class. Quantum Grav.* **27** 025006 (*Preprint* 0906.4787)
[18] Gubser S S, Klebanov I R and Polyakov A M 1998 *Phys. Lett.* B **428** 105 (*Preprint* hep-th/9802109)
[19] Witten E 1998 *Adv. Theor. Math. Phys.* **2** 253 (*Preprint* hep-th/9802150)
[20] Neupane I P and Dadhich N 2009 *Class. Quantum Grav.* **26** 015013 (*Preprint* 0808.1919)
[21] Ciobanu T and Ramirez D M 2017 *Preprint* 1708.04997
[22] Myers R C and Simon J Z 1988 *Phys. Rev.* D **38** 2434
[23] Cai R G 2002 *Phys. Rev.* D **65** 084014 (*Preprint* hep-th/0109133)
[24] Cai R G 2004 *Phys. Lett.* B **582** 237 (*Preprint* hep-th/0311240)
[25] Ge X H, Matsuo Y, Shu F W, Sin S J and Tsukioka T 2008 *J. High Energy Phys.* JHEP10(2008)009 (*Preprint* 0808.2354)
[26] Ge X H and Sin S J 2009 *J. High Energy Phys.* JHEP05(2009)051 (*Preprint* 0903.2527)
[27] Chamblin A, Emparan R, Johnson C V and Myers R C 1999 *Phys. Rev.* D **60** 064018 (*Preprint* hep-th/9902170)
[28] Cvetič M, Nojiri S and Odintsov S D 2002 *Nucl. Phys.* B **628** 295 (*Preprint* hep-th/0112045)
[29] He S, Li L F and Zeng X X 2017 *Nucl. Phys.* B **915** 243 (*Preprint* 1608.04208)
[30] Hu C, Zeng X and Liu X 2013 *Sci. China Phys. Mech. Astron.* **56** 1652
[31] Gibbons G and Hartnoll S A 2002 *Phys. Rev.* D **66** 064024 (*Preprint* hep-th/0206202)
[32] Dotti G and Gleiser R J 2005 *Phys. Rev.* D **72** 044018 (*Preprint* gr-qc/0503117)
[33] Lucas A 2015 *J. High Energy Phys.* JHEP03(2015)071 (*Preprint* 1501.05656)
[34] Nguyen C T, Desgranges F, Galanis N, Roy G, Maré T, Boucher S and Angue Mintsa H 2008 *Int. J. Therm. Sci.* **47** 103
[35] Knittel D R, Pack S P, Lin S H and Eyring L 1977 *J. Chem. Phys.* **67** 134
[36] Bertotti G 1998 *Hysteresis in Magnetism: For Physicists, Materials Scientists, and Engineers* (San Diego, CA, USA: Academic Press)

IOP Conf. Series: Journal of Physics: Conf. Series **942** (2017) 012019 doi:10.1088/1742-6596/942/1/012019

# Noncommutative Black Holes at the LHC

**Elena Michelle Villhauer**

University of California, Irvine; ATLAS, CERN

E-mail: elena.michelle.villhauer@cern.ch

**Abstract.** Based on the latest public results, 13 TeV data from the Large Hadron Collider at CERN has not indicated any evidence of hitherto tested models of quantum black holes, semiclassical black holes, or string balls. Such models have predicted signatures of particles with high transverse momenta. Noncommutative black holes remain an untested model of TeV-scale gravity that offers the starkly different signature of particles with relatively low transverse momenta. Considerations for a search for charged noncommutative black holes using the ATLAS detector will be discussed.

## 1. Introduction to the Hierarchy Problem and Black Hole Formation at the LHC

The European Organization of Nuclear Research, CERN, seeks to study the Standard Model of particle physics, its problems, and possible solutions. Arguably the most outstanding and pressing problem that the Standard Model fails to address is the irreconcilability of general relativity and quantum mechanics. Quantum gravity is a theory that attempts to explain gravity according to the principles of quantum mechanics. The quantum effects of gravity are believed to become strong and apparent at the Planck scale, which, in four dimensions, is estimated to be around 1 x $10^{19}$ GeV. Another prominent and pressing problem that the Standard Model fails to address is the hierarchy problem, or the extreme discrepancy between aspects of gravity and the weak force. Out of the four fundamental forces, the gravitational force is by far the weakest, ranking approximately 32 orders of magnitude weaker than the weak force, 38 orders of magnitude weaker than the strong force, and 36 orders of magnitude weaker than the electromagnetic force. As noted by Arkani-Hamed, S. Dimopoulos and G. Dvali, one way to account for the extreme discrepancy in gravity's relative strength, the hierarchy problem, is the existence of extra dimensions [3, 4]. Gravity, unlike the other fundamental forces, may propagate in more than four dimensions, thereby reducing its apparent strength. With more than four dimensions, the Planck scale may be lowered to the low TeV-scale, the energy range accessible by the Large Hadron Collider in Geneva, Switzerland [4]. This allows for the possibility of black hole production. A black hole is predicted to be formed at the LHC if, as proposed by Thorne, the hoop conjecture is fulfilled by two partons from a proton-proton collision [9, 16]. The classical Hoop Conjecture states that a black hole can be formed if an imploding object is compressed in all directions into a region with a circumference smaller than $2\pi$ times the object's Schwarzschild radius [16]. While LHC black hole event generators incorporate the classical hoop conjecture, a more precise and accurate description of the hoop conjecture for quantum black hole production at the LHC can be found in a study of the quantum hoop conjecture by Casadio, Micu, and Scardigli [9]. The main types of black hole and signatures of TeV-scale predicted to

IOP Conf. Series: Journal of Physics: Conf. Series **942** (2017) 012019 doi:10.1088/1742-6596/942/1/012019

be formed at the LHC are semiclassical black holes, quantum black holes, and string balls. The latest results for searches for each type of black hole formation will follow.

## 2. Models Tested and 13 TeV Results

### *2.1. Semiclassical Black Holes*

Semiclassical black holes at the LHC are primarily modeled by ATLAS and CMS with the Charybdis2 and BlackMax generators using ADD-type extra dimensions. Two ATLAS semiclassical black hole searches have been performed with 13 TeV data: a multijet search and a lepton plus jets search. For the multijet search, the selection included at least three jets with scalar sum of jet transverse momenta greater than 1 TeV. Using 13 TeV data with an integrated luminosity of 3.6 $fb^{-1}$, the ATLAS multijet search set exclusion limits for rotating black holes in 6 extra dimensions with mimumum black hole masses of 9.0 TeV - 9.7 TeV [7]. The lepton plus jet search set a selection consisting of at least three objects: a leading lepton with pT $>$ 100 GeV and at least two other objects (jets or leptons) with pT $>$ 100 GeV. Additionally, the $\sum$ pT was required to be $>$ 2 TeV or 3 TeV. The lepton plus jet search used data with an integrated luminosity of 3.2 $fb^{-1}$ to exclude rotating black holes in two extra dimensions with minimum black hole masses up to 7.8 TeV for MD = 2 TeV and with minimum black hole masses of 7.4 TeV for MD = 5 TeV [8]. CMS undertook a search for semiclassical black holes decaying to multijets, in which data of an integrated luminosity of 2.3 $fb^{-1}$ was used to exclude semiclassical black holes with masses as high as 9.5 TeV [10].

### *2.2. Quantum Black Holes*

The ATLAS experiment has set limits on the following QBH decay channels using 13 TeV data: QBH $\rightarrow$ dijet, QBH $\rightarrow$ photon + jet. The ATLAS dijet search used data with 37 $fb^{-1}$ to set a 95 percent confidence level exclusion on quantum black holes up to masses of 8.9 TeV [5]. The ATLAS photon + jet search used data with an integrated luminosity of 36.7 $fb^{-1}$ to exclude QBH in RS-type extra dimensions below 4.4 TeV and QBH in ADD-type extra dimensions below 7.1 TeV [6]. The CMS experiment has used 13 TeV data with an integrated luminosity of 3 $fb^{-1}$ to set limits on QBH $\rightarrow$ dijet for 6 dimensional ADD-type QBH with a lower limit mass of 7.8 TeV and for 5 dimensional RS-type QBH with a lower limit mass of 5.3 TeV. The CMS multijet used 13 TeV data with an integrated luminosity of 2.3 $fb^{-1}$ to exclude quantum black hole masses as high as 9.0 TeV [10].

### *2.3. String Balls*

ATLAS searches for stringballs in the multijet final state [7] using data 3.0 $fb^{-1}$ to exclude string balls as high as 9 TeV. CMS used data of an integrated luminosity of 2.3 $fb^{-1}$ to exclude string balls with masses as high as 9.5 TeV [10].

## 3. Results Interpretation

### *3.1. Reasons why extra dimensions have not yet been discovered*

ATLAS and CMS have used 13 TV data from the 2015-2017 Run II campaign to search for semiclassical black holes, quantum black holes, and string balls. Based on the latest public results, no evidence of such models has been found and exclusion limits have been set. The absence of evidence for signatures of TeV-scale gravity at the LHC can be attributed to either one of three reasons. It could be that extra dimensions do not exist and any subsequent search for signatures of extra dimensions will prove futile. However, it could also be that extra dimensions do exist, and that we have not discovered evidence of them because we need more energy to access them. Finally, and most excitingly, signatures of TeV-scale gravity may have already been produced at the LHC, but have either been hiding in signatures different than those we

IOP Conf. Series: Journal of Physics: Conf. Series **942** (2017) 012019 doi:10.1088/1742-6596/942/1/012019

have used in searches or because they have been produced at a very low rate and the signal is diluted.

### *3.2. The LHC's future plans and the best bet for TeV-scale gravity*

The LHC has been producing collisions at 13 TeV since April 2015. However, at the end of 2018, the LHC plans to enter a long shutdown, at which point all data taking will stop. Collisions are scheduled to resume at the start of 2020 with energies of 14 TeV. This is the only planned energy increase in the set future of the LHC. All other improvements scheduled until 2035 will be regarding luminosity, as part of the High Luminosity Large Hadron Collider(the HL-LHC), the version of the LHC intended for luminosity upgrades up to 3000 $\text{fb}^{-1}$ [2]. This poses a problem for models of TeV-scale gravity already tested because black hole cross sections increase exponentially with increases in energy, but only linearly with increases in luminosity. Upon increasing the energy of the LHC from 8 TeV to 13 TeV, the 13 TeV to 8 TeV cross section ratio of quantum black holes with masses of 6 TeV becomes 9000. When increasing from 13 TeV to 14 TeV, the 14 TeV to 13 TeV cross section ratio of quantum black holes with masses of 9 TeV in 6 extra dimensions becomes 5.3 TeV. Larger increases in energy, therefore, would provide the best opportunity for discovery. However, increases in energy will not occur until after the long shutdown at the winter of 2018, after which energy is set to increase to 14 TeV. There is perhaps a possibility to increase energies by replacing a magnet. However, no such plans have been set for the near future. Therefore, the best bet for finding new physics with black hole remains looking at different models yielding different signatures. Searches for semiclassical black holes, quantum black holes, and other signatures of TeV-scale gravity at the LHC have all targeted final states containing high transverse momenta. Noncommutative black holes offer a very different experimental signature from other potential signatures of TeV-scale gravity: many particles with a low transverse momenta spectra.

## 4. Noncommutative Black Holes

As described by Nicolini [14], noncommutative black holes are predicted to form in models where noncommutative geometry is incorporated into the framework of extra dimensions. The defining feature of noncommutativity in black holes is that the spacetime coordinates $\hat{x}^A$ and $\hat{x}^B$ do not commute [14], such that

$$\left[\hat{x}^A, \hat{x}^B\right] = i\theta^{AB} \equiv i\frac{\epsilon^{AB}}{\Lambda_{\text{NC}}^2}\,, \tag{1}$$

where, taking directly from Gingrich [11], "$\theta^{AB}$ is an real antisymmetric $D \times D$ matrix. For convenience, we have separated the mass scale $\Lambda_{\text{NC}}$ associated with noncommutative from the dimensionless matrix structure $\epsilon^{AB}$ of $\theta^{AB}$. If $\Lambda_{\text{NC}}^{-2}$ is the average magnitude of the elements in $\theta^{AB}$, we assume the elements of $\epsilon^{AB}$ are of $\mathcal{O}(1)$."

Noncommutativity invokes a smearing in the mass distribution [1] such that the charge density [15], $\rho$, becomes [11]

$$\rho = \frac{m}{(4\pi\theta)^{(n+3)/2}} e^{-r^2/(4\theta)}\,, \tag{2}$$

where $\sqrt{\theta} = 1/\Lambda_{\text{NC}}$

Noncommutative black holes are theoretically of interest due to the resulting nonsingular solution, but most importantly, because they form an effective theory to quantum gravity. Experimentally, this effective theory allows for a new mass threshold above the Planck scale and, thus, a new energy regime for physics beyond the Standard Model [11]. Noncommutative black holes are cold and decay to a stable remnant [1], [14]. The opportunity to study stable remnants through noncommutative black hole studies is exciting because stable remnants appear in a variety of models ranging from those of loop quantum gravity and string gravity to tunneling [11].

IOP Conf. Series: Journal of Physics: Conf. Series **942** (2017) 012019 doi:10.1088/1742-6596/942/1/012019

## 5. 2010 Neutral Noncommutative Black Hole Study

A study undertaken in 2010 by Gingrich [11] shows the phenomenological aspects of noncommutative black holes at the LHC. The cross sections and temperatures for noncommutative black holes, shown in Figure 2 [11], are significantly lower than those of other signatures of TeV-scale gravity. The combined effect is that noncommutative black holes yield softer particles than semiclassical black holes, string balls, or quantum black holes.

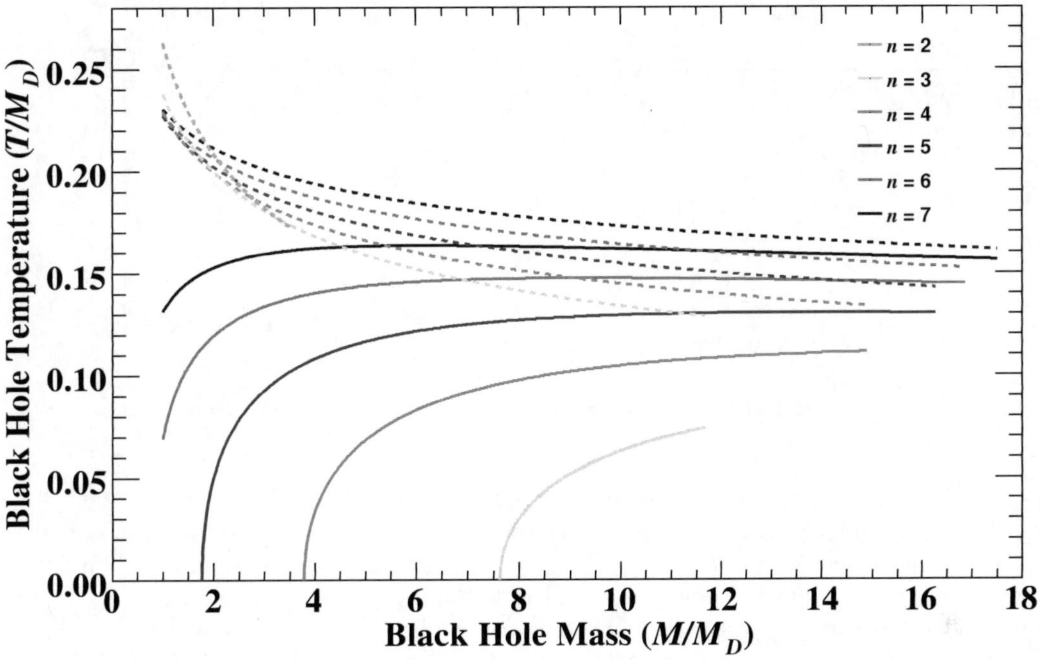


**Figure 1.** Black Hole Temperature vs. Black Hole Mass (M/$M_D$). Solid lines: NC BH, Dashed lines: Commutative BH. Figure taken from [11].

There is an average of 9 primary decay particles from the black hole and a maximum of 25. The average transverse momentum of the final state particles is 70 GeV. A generator cut-off was imposed at 100 MeV above the mass of the remnant due to technical complications regarding the generator's efficiency near the mass of the remnant [11]. Without this cutoff, the primary decay particle multiplicity would be greater and the average transverse momentum of final state particles would be even lower. The majority of events in the simulations of the study contain at least one jet, while about 45 percent have a leading lepton. Figure 2 [11] depicts the soft jet maximum transverse momentum and the soft lepton maximum transverse momenta. This soft transverse momentum spectrum of the final state particles allows for the noncommutative black hole signal to blend into QCD background. Searches for semiclassical black holes with low masses set a selection on scalar sum of transverse momentum of greater than a few TeV. Due to the cold nature of noncommutative black holes, the scalar sum of transverse momentum can reach zero. This eliminates the ability to include scalar sum of transverse momentum in the selection [11].

As for the characteristics of the noncommutative black hole's stable remnant, the remnant is slow, with the most likely speed being .3c, and has an average transverse momentum of less than 230 GeV.

IOP Conf. Series: Journal of Physics: Conf. Series **942** (2017) 012019 doi:10.1088/1742-6596/942/1/012019

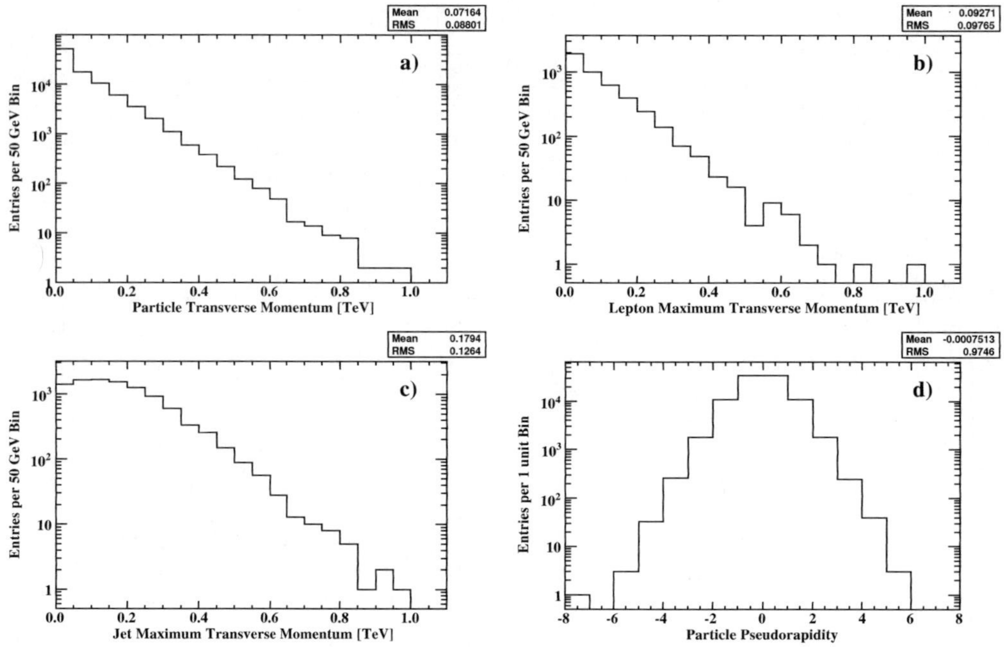


**Figure 2.** Characteristics of Final State Particles. Figure taken from [11].

## 6. Strategies for a Charged Noncommutative Black Hole Search

The 2010 LHC noncommutative black hole studies by Gingrich [11] show that noncommutative black holes offer the unique signature of particles with low transverse momentum. This was the only study of noncommutative black holes using LHC data, and it was stopped at trigger level due to the low transverse momentum spectrum of the final state particles being obscured by the QCD background. This study was undertaken for a neutral noncommutative black hole model. The addition of charge would further lower the transverse momentum spectrum, leading to a more dramatic signature. Therefore, a search for charged noncommutative black holes, in which the charge of the black hole is confined to the brane, is intended. The following is a discussion regarding possible search options and ways to separate the low pT signal from the QCD background.

A first step would be to look at the HT for each noncommutative black hole event and compare it to that of QCD. Thorough studies should be undertaken to compare the shape of black hole radiation to that of QCD. It might be useful to check if the black hole thermal distribution matches the KNO distribution. The signal may be able to be extracted from the QCD background by treating the signal and background as multijets and studying the isotropicness of the noncommutative black hole versus the isotropicness of the QCD background.

Adding a W or Z to the selection could help reduce the background, but comes at the risk of chopping the probability by a factor of alpha. For charged noncommutative black holes, adding a photon in the selection could help separate the signal from background. A search for multijets plus photons has not been done before, so this would have the added benefit of exploring a new final state.

The study by Gingrich points out that another way to detect noncommutative black holes is to search for the remnant. If the remnant is charged, once can use the fact that the remnant is slow to search for charged, long lived particles. Additionally, for a charged remnant, there should be a balancing charge. Perhaps this balancing charge looks like a charged lepton combiation, which can be used as a selection requirement.

## 7. Conclusion

Based on the latest public results, no evidence of hitherto tested models of quantum black holes, semiclassical black holes, or string balls have been found at the LHC using 13 TeV data. Noncommutative black holes offer a chance to use a different signature to search for TeV-scale gravity. A search is intended for charged noncommutative black holes using the ATLAS detector

## Acknowledgments

The author wishes to thank Marco Sampaio for the crucial specification that, for a charged black hole model to be incorporated in Charybdis2, the charge of the black hole must be confined to the brane. Additionally, the author would like to thank Frank Krauss, Richard Keith Ellis, and Daniel Maitre for their keen insight on separating low pT signal from QCD background, which was subsequently included above.

## References

[1] Ansoldi, Nicolini, Smailagic, Spallucci, *Noncommutative geometry inspired charged black holes*, Phys.Lett.B645:261-266,2007. arXiv:gr-qc/0612035.
[2] G. Apollinari, O. Bruening, T. Nakamoto, L. Rossi, *High Luminosity Large Hadron Collider HL-LHC.* arXiv:1705.08830.
[3] N. Arkani-Hamed, S. Dimopoulos and G. Dvali, *Phenomenology, Astrophysics and Cosmology of Theories with Sub-Millimeter Dimensions and TeV Scale Quantum Gravity*, Phys.Rev.D59 (1999) 086004, arXiv:hep-ph/9807344.
[4] Nima Arkani-Hamed, Savas Dimopoulos, Gia Dvali, *The Hierarchy Problem and New Dimensions at a Millimeter*, Phys.Lett.B429:263-272,1998, arXiv:hep-ph/9803315.
[5] ATLAS Collaboration, *Search for new phenomena in dijet events using 37 $fb^{-1}$ of pp collision data collected at $s = \sqrt{13}$ with the ATLAS detector*
[6] ATLAS Collaboration, *Search for new phenomena in high-mass final states with a photon and a jet from pp collisions at $s = \sqrt{13}$ with the ATLAS detector*
[7] ATLAS Collaboration, *Search for strong gravity in multijet final states produced in pp collisions at $s = \sqrt{13}$ using the ATLAS detector at the LHC*, arXiv:1512.02586, 2016.
[8] ATLAS Collaboration, *Search for TeV-scale gravity signatures in high-mass final states with leptons and jets with the ATLAS detector at $s = \sqrt{13}$*, Physics Letters B 760 (2016) 520-537
[9] R. Casadio, O. Micu and F. Scardigli, *Quantum hoop conjecture: Black hole formation by particle collisions*, Phys. Lett. B 732, 105 (2014).
[10] CMS Collaboration, *Search for black holes in high-multiplicity final states in proton-proton collisions at $s = \sqrt{13}$*, arXiv:1705.01403, May 2017.
[11] Doug Gingrich, *Noncommutative geometry inspired black holes in higher dimensions at the LHC*, JHEP05(2010)022, arXiv:1003.1798[hep-ph].
[12] Douglas M. Gingrich, *Quantum black holes with charge, colour, and spin at the LHC*, J.Phys.G37:105108,2010, arXiv:0912.0826 [hep-ph].
[13] Douglas M. Gingrich and Kevin Martell, *Study of highly-excited string states at the Large Hadron Collider*, Phys.Rev.D78:115009,2008, arXiv:0808.2512 [hep-ph].
[14] Piero Nicolini, Noncommutative Black Holes, The Final Appeal To Quantum Gravity: A Review, Int.J.Mod.Phys.A24:1229-1308,2009, arXiv:0807.1939[hep-th]
[15] Thomas G. Rizzo, *Noncommutative Inspired Black Holes in Extra Dimensions*, JHEP 0609 (2006) 021, arXiv:hep-ph/0606051.
[16] K. S. Thorne, *Nonspherical Gravitational Collapse: A Short Review*, Magic Without Magic, San Francisco 1972, 231-258.

IOP Conf. Series: Journal of Physics: Conf. Series **942** (2017) 012020 doi:10.1088/1742-6596/942/1/012020

# AdS/CFT far from equilibrium in a Vaidya setup

**Michael F Wondrak**[1,2], **Matthias Kaminski**[3], **Piero Nicolini**[1,2] **and Marcus Bleicher**[1,2]

[1] Frankfurt Institute for Advanced Studies (FIAS),
Ruth-Moufang-Str. 1, 60438 Frankfurt am Main, Germany
[2] Institut für Theoretische Physik, Johann Wolfgang Goethe-Universität Frankfurt am Main,
Max-von-Laue-Str. 1, 60438 Frankfurt am Main, Germany
[3] Department of Physics and Astronomy, University of Alabama,
514 University Blvd., Tuscaloosa, AL 35487-0324, USA

E-mail: wondrak@fias.uni-frankfurt.de, mski@ua.edu,
nicolini@fias.uni-frankfurt.de, bleicher@fias.uni-frankfurt.de

**Abstract.** In this paper we apply the AdS/CFT correspondence to study a strongly coupled plasma far from equilibrium with a strong emphasis on the shear behavior. The plasma serves as a model for an electrically charged quark-gluon plasma. On the gravitational side, we use an ingoing Vaidya black brane spacetime. The highest rate of mass infall is confined to a short time interval.

## 1. Introduction

According to present cosmological models, the universe was radiation-dominated after the inflation phase and while expanding it cooled down following the law $T \propto t^{-1/2}$ where $T$ is the temperature and $t$ is the time after the big bang. Until about $10^{-5}$ s the temperature exceeded a value of 200 MeV and the universe was filled with a very special state of matter called the quark-gluon plasma (QGP) [1]. Such a state is supposed to be present in the interior of neutron stars and in supernova explosions. The QGP describes highly compressed baryonic matter at high temperatures subject to quantum chromodynamics (QCD). QCD has a large coupling constant at low energies which implies that the fundamental degrees of freedom, quarks and gluons, do not manifest individually, but are confined to hadrons. At larger energies corresponding to temperatures close to the critical temperature $T_c \sim 200$ MeV, the coupling decreases so that quarks and gluons deconfine and form the QGP. Finally, at even higher energies, the theory becomes weakly coupled.

The treatment of the QGP around $T_c$ is challenging because the QCD coupling constant is too large to allow for perturbation theory. There is much activity to describe the QGP by, *e.g.*, hydrodynamics, kinetic theory, effective field theory, and lattice QCD (for a detailed treatment cf. *e.g.* [2]). A particular approach is provided by the AdS/CFT correspondence. It connects a supergravity theory on an asymptotically Anti-de Sitter (AdS) spacetime with a non-abelian conformal field theory (CFT) on the flat conformal boundary of that spacetime. As a strong-weak duality, the gravitational spacetime is weakly curved, while the conformal field theory is strongly coupled. Even more general versions have been conjectured, the strong and strongest form, which also allow stronger curvature on the gravity side and smaller couplings on the field

IOP Conf. Series: Journal of Physics: Conf. Series **942** (2017) 012020 doi:10.1088/1742-6596/942/1/012020

theory side. Furthermore, the gauge/gravity duality can also address non-conformal fields on the boundary (cf. *e.g.* [3]).

On the experimental side, a QGP has already been produced in heavy-ion collisions at the Relativistic Heavy Ion Collider (RHIC) and the Large Hadron Collider (LHC) which was proven by observation of elliptic flow and jet quenching [4, 5]. Experimental data allows to extract how the highly inhomogeneous initial state after the collision comes to local equilibrium and how it reacts to fluctuations. The corresponding transport coefficients serve as benchmarks for theoretical approaches towards the QGP. One prominent representative is the ratio of shear viscosity to entropy density, $\eta/s$. Its extrapolated value lies close to the AdS/CFT prediction of $1/(4\pi)$ for gauge theories at large 't Hooft coupling which possess a supergavity dual [6–8].

While the hydrodynamic description requires local thermodynamic equilibrium and is therefore applicable only at advanced times after the heavy-ion collision, the AdS/CFT correspondence can be used at earlier times in the far-from-equilibrium regime [9–14].

We focus our interest on determining the shear behavior of a hot, charged, strongly coupled plasma far from equilibrium. The paper is organized as follows: In section 2 we introduce the quantities relevant for our calculation and show relations among them based on the AdS/CFT correspondence. Moreover, we discuss shear for long-scaled perturbations in the hydrodynamic regime. A sketch of our calculation procedure together with the spacetime model can be found in section 3. Section 4 focuses on the evolution of fluctuations using numerical methods. We conclude in section 5.

## 2. AdS/CFT correspondence and hydrodynamics

The AdS/CFT correspondence allows to investigate a quantum field theory at finite temperature by studying a black brane, *i.e.*, an infinitely extended black hole with planar topology in a higher-dimensional product spacetime which consists of an asymptotically AdS spacetime and a compact manifold [15]. The field can be thought to be located on the $d$-dimensional boundary of the $(d+1)$-dimensional AdS bulk spacetime. Since it is possible to associate thermodynamic properties to black objects (black hole thermodynamics), several quantities can be compared directly. For example, the temperature of the field on the boundary, $T$, is given by the Hawking temperature of the black brane, $T_{\mathrm{H}}$. The entropy of the field state equals the Bekenstein-Hawking entropy, $S_{\mathrm{BH}}$. Therefore the entropy density $s$ follows from the ratio between $S_{\mathrm{BH}}$ and the field extension, which is $\mathrm{Vol}(\mathbb{R}^{d-1})$ for the Poincaré patch:

$$T = T_{\mathrm{H}} = \frac{\kappa}{2\pi}, \qquad s = \frac{S_{\mathrm{BH}}}{\mathrm{Vol}(\mathbb{R}^{d-1})} = \frac{A}{4G\,\mathrm{Vol}(\mathbb{R}^{d-1})}. \tag{1}$$

Here $\kappa$ and $A$ denote the surface gravity and the horizon area of the black brane and $G$ is the higher-dimensional gravitational constant.

The AdS/CFT correspondence or, more generally, the gauge/gravity duality states the equality of the grand-canonical partition function of the field theory on the boundary at $r \to \infty$ and that of supergravity in the bulk. The GKP-Witten relation dynamically connects the field quantities, *e.g.*, a U(1) current on the boundary with a U(1) bulk field. By choosing a non-vanishing electric chemical potential $\mu$ on the boundary, the field state becomes electrically charged. The electrodynamic U(1) gauge symmetry on the boundary is modeled by the trace of the maximal abelian subgroup of the SU($\mathcal{N}$) R-symmetry where $\mathcal{N}$ is the number of supercharges present [3]. The bulk gauge field $A_m$ gives rise to the chemical potential by its boundary asymptotic behavior,

$$\mu = A_0|_{r_{\mathrm{h}}} - A_0|_{r\to\infty} \tag{2}$$

where $r_{\mathrm{h}}$ denotes the position of the black brane apparent horizon. The energy-momentum tensor of the field theory is sourced by the boundary value of the bulk metric $g_{mn}$. Since

IOP Conf. Series: Journal of Physics: Conf. Series **942** (2017) 012020 doi:10.1088/1742-6596/942/1/012020

shear introduces transverse momentum diffusion, the field response is encoded in the spatial off-diagonal components of the energy-momentum tensor. The corresponding metric components are of particular interest for us.

The quark-gluon plasma in the experimentally accessible regime behaves as a strongly coupled fluid. This is due to the fact that the energies in heavy-ion collisions are still too small to decrease the coupling constant significantly. Close to equilibrium, the QGP evolution and its reaction to perturbations long-scaled in space and time can be described by relativistic hydrodynamics. In contrast to an ideal fluid (IF) which is inviscid, fluctuations in the QGP are damped. In order to capture the dissipative behavior, the constituent equation for the ideal fluid energy-momentum tensor is replaced by a derivative expansion. At zeroth order there are the components of an ideal fluid, $T^{\rm IF}_{\mu\nu}$, while the non-ideal fluid (NIF) components are contained in $T^{\rm NIF}_{\mu\nu}$ [16, 17]. Up to first order we obtain

$$\begin{aligned} T_{\mu\nu} &= T^{\rm IF}_{\mu\nu} + T^{\rm NIF}_{\mu\nu} \\ &= (\epsilon\, u_\mu u_\nu + P\,\Delta_{\mu\nu}) + \left(-\eta\,\sigma_{\mu\nu} - \zeta\,\nabla_\rho u^\rho\,\Delta_{\mu\nu} + \mathcal{O}(\nabla^2)\right). \end{aligned} \tag{3}$$

Referring to a given reference frame, the fluid velocity is denoted by $u^\mu$, the energy density by $\epsilon$, and the pressure by $P$. The projector $\Delta_{\mu\nu}$ and the shear tensor $\sigma_{\mu\nu}$ read

$$\Delta_{\mu\nu} \equiv g_{\mu\nu} + u_\mu u_\nu, \qquad \sigma_{\mu\nu} \equiv 2\nabla_{\langle\mu}u_{\nu\rangle} \tag{4}$$

where angle brackets stand for the the symmetric traceless part of a tensor with respect to the corresponding indices. The shear viscosity $\eta$ and bulk viscosity $\zeta$ are examples of transport coefficients. The latter describes the frictional resistance to compression or expansion. The former determines the transverse momentum diffusion which manifests itself, *e.g.*, as the velocity profile of a fluid when applying shear. Consider a 3-dimensional case with a local Cartesian rest frame and a fluid between two surfaces parallel to the $x$-$z$ plane. A force in the $x$-direction applied tangentially to one surface with an amount $F$ per area $A$ leads to a velocity gradient $\partial_y u_x$,

$$-T_{yx} = \frac{F}{A} = \eta\,\sigma_{yx} = \eta\,\partial_y u_x. \tag{5}$$

Thus the shear viscosity corresponds to a non-vanishing off-diagonal transverse component of the energy-momentum tensor.

## 3. Calculation procedure

As we have seen in the preceding section, shear is related to the $xy$-component of the energy-momentum tensor $T^{\mu\nu}$. For field states close to equilibrium the shear viscosity $\eta$ can be obtained by a Green-Kubo relation from the associated retarded Green's function $G^{xy,xy}_{\rm R}$.

We are interested in the behavior of the quark-gluon plasma under shear in the far-from-equilibrium regime and thus outside of the hydrodynamic range of applicability. This is the sketch of our approach: In the same way as above, we focus on $T^{xy}$ and apply linear response theory. We introduce a linearized metric perturbation, $h_{xy}$, in order to obtain the linearized answer in the expectation value of the energy-momentum tensor, $\langle\delta T^{xy}\rangle$. Since in our setup the corresponding background quantities vanish, $g^{\rm (bg)}_{xy} = 0$, $\left\langle T^{{\rm (bg)}\,xy}\right\rangle = 0$, we have $g_{xy} = h_{xy}$ and $\langle T^{xy}\rangle = \langle\delta T^{xy}\rangle$.

The field state far from equilibrium is achieved by choosing a rapidly changing background solution in the bulk. Therefore time-translation invariance is broken, but the background is still spatially homogeneous and isotropic. This allows us to perform a Fourier transformation of the metric perturbation with respect to the transverse coordinates $\vec{x}$ to get rid of the corresponding derivatives. The equation of motion of the metric fluctuation is solved numerically with the infalling boundary condition at the black brane horizon. The one-point function $\langle T^{xy}\rangle$ is extracted from the boundary behavior of the bulk field. Choosing a special profile for the metric fluctuation, a Dirac delta distribution, the one-point function in presence of this source is the retarded Green's function. This allows us to study the two-point function in time [11, 13, 14].

In our investigation, we want to drive a homogeneous and isotropic holographic plasma far from equilibrium by a sudden temperature rise. Therefore, we choose a 4-dimensional asymptotically AdS Vaidya spacetime in the bulk (cf. *e.g.* [18, 19]; for a related study on this subject cf. [20]). The metric in Poincaré form reads

$$\mathrm{d}s^2 = r^2\left[-\left(1-\frac{M(v)}{r^3}+\frac{Q^2(v)}{2r^4}\right)\mathrm{d}v^2+\mathrm{d}x^2+\mathrm{d}y^2+\frac{2}{r^2}\,\mathrm{d}r\,\mathrm{d}v\right]. \tag{6}$$

where $v$ is the ingoing Eddington-Finkelstein coordinate which coincides with ordinary time $t$ on the boundary. The AdS radial coordinate is denoted by $r$, and the coordinates $x$ and $y$ denote Cartesian non-compact transverse coordinates. The Vaidya solution generalizes the Schwarzschild black brane to the case of a time-dependent mass $M(v)$. We further allow for a time-dependent electric charge $Q(v)$. The apparent horizon of the black brane is located at $r_{\mathrm{h}}(v)$. Here and in the following, all quantities are normalized to appropriate powers of the AdS radius $L$.

The time dependence is due to the infall of pressureless matter which propagates at the speed of light, known as null matter. By choosing the mass of the brane to increase rapidly in a short null time interval, we provide the intended energy deposition. This is realized by the mass profile

$$M(v) = m + \frac{1}{2}\,m_{\mathrm{s}}\left(1+\tanh\left(\frac{v-t_{\mathrm{s}}}{\sigma_{\mathrm{s}}}\right)\right) \tag{7}$$

where the steepness is parametrized by $\sigma_{\mathrm{s}}$. The initial mass is denoted by $m$, the infalling mass by $m_{\mathrm{s}}$. At time $t_{\mathrm{s}}$ on the boundary, the temperature undergoes the phase of maximum increase rate. An example of this profile is shown in Fig. 1. The infalling null matter contributes to the

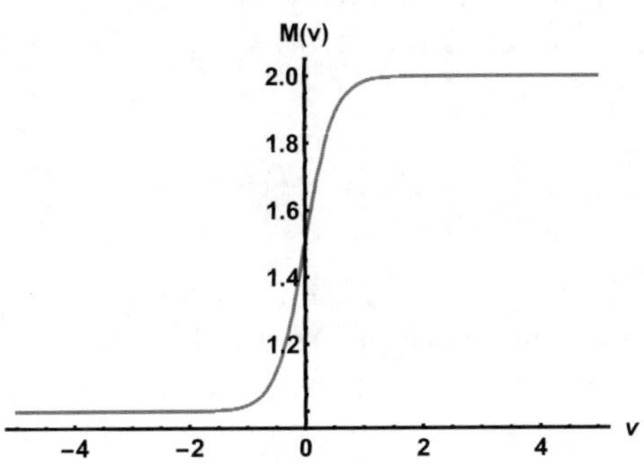


**Figure 1.** Sample mass profile $M(v)$ of the black brane due to the infall of null matter. It shows the smooth transition from the initial mass $m = 1$ to the final mass $m + m_{\mathrm{s}} = 2$ centered around $t_{\mathrm{s}} = 0$ within a short time period determined by $\sigma_{\mathrm{s}} = 0.5$.

bulk energy-momentum tensor and carries an electric current,

$$T^{\text{bulk, null}}_{mn} = \frac{m_{\mathrm{s}}\,\mathrm{sech}^2\left(\frac{t_{\mathrm{s}}-v}{\sigma_{\mathrm{s}}}\right)\,r - 2\sigma_{\mathrm{s}}\,Q(v)\,\partial_v Q(v)}{2\sigma_{\mathrm{s}}\,r^3}\,\delta^v_m\,\delta^v_n, \qquad J^m_{\text{bulk, null}} = -\frac{\partial_v Q(v)}{r^2}\,\delta^m_r. \tag{8}$$

Furthermore, the black brane has a time-dependent electric charge $Q(v)$ and couples to the U(1) gauge field $A_m$. The gauge field takes on a non-trivial value at the boundary which plays the role of the electric chemical potential $\mu(v)$ on the field theory side,

$$A_m(v,r) = -\left(\mu(v)+\frac{Q(v)}{r}\right)\delta^v_m, \qquad \mu(v) = -\frac{Q(v)}{r_{\mathrm{h}}(v)}. \tag{9}$$

## 4. Metric fluctuation

To study the properties of the plasma, we introduce linearized perturbations of the background fields, namely $h_{\mu\nu}$ associated to the metric and $a_\mu$ associated to the gauge field. We apply the radial gauge in which the $r$-components of the fluctuations vanish. In the linearized equations of motion, $h_{xy}$ decouples from the other fluctuations. Using the inverse radial coordinate $z \equiv 1/r$ and rescaled metric fluctuation $\hat{h}_{xy} \equiv h_{xy}/z^2$, the equation of motion reads

$$\begin{aligned} 0 &= -2z\left(1 - M(v)\, z^3 + \frac{Q^2(v)}{2}\, z^4\right) \partial_z^2 \hat{h}_{xy} \\ &\quad + \left(4 + 2M(v)\, z^3 - 2Q^2(v)\, z^4\right) \partial_z \hat{h}_{xy} - 4\,\partial_v \hat{h}_{xy} + 4z\,\partial_v \partial_z \hat{h}_{xy}. \end{aligned} \tag{10}$$

The equation of motion can be solved numerically. The analytic boundary expansion of the field in terms of the radial coordinate reveals two independent coefficients, the source $h_{xy}^{(0)}$ of the boundary operator $T^{xy}$, as well as, at subleading order, the expectation value $\langle T^{xy}\rangle$ up to contact terms. According to linear response theory, we can obtain the Green's function in position space by evaluating the one-point function in presence of a Dirac delta-like source $\hat{h}_{xy}^{\text{pulse}}$,

$$\begin{aligned} \langle T^{xy}(t_1, \vec{x}_1)\rangle_h &= -\int \mathrm{d}^d\xi\; G_{\mathrm{R}}^{xy,xy}\left(t_1, \tau; \vec{x}_1 - \vec{\xi}\right) \underbrace{\hat{h}_{xy}^{\text{pulse}}(\xi)}_{=\delta(\tau - t_2)\,\delta(\vec{\xi} - \vec{x}_2)} \\ &= -G_{\mathrm{R}}^{xy,xy}(t_1, t_2; \vec{x}_1 - \vec{x}_2)\,. \end{aligned} \tag{11}$$

A sketch of the spacetime including the metric fluctuation is presented in Fig. 2. In order to implement the boundary source in the numerical solution procedure, we approximate it by a Gaussian profile,

$$\hat{h}_{xy}^{\text{pulse}}(v) = A_{\mathrm{p}}\,\delta(v - t_{\mathrm{p}}) \simeq \frac{A_{\mathrm{p}}}{\sqrt{2\pi}\,\sigma_{\mathrm{pv}}}\,\exp\left(-\frac{1}{2}\left(\frac{v - t_{\mathrm{p}}}{\sigma_{\mathrm{pv}}}\right)^2\right). \tag{12}$$

Here $A_{\mathrm{p}}$ denotes the integrated amplitude of the pulse with width $\sigma_{\mathrm{pv}}$ at boundary time $t_{\mathrm{p}}$. In this way, we can deduce the retarded Green's function from the boundary behavior of the fluctuation in an easy way. Currently, the results of this numerical calculation are being checked and analyzed.

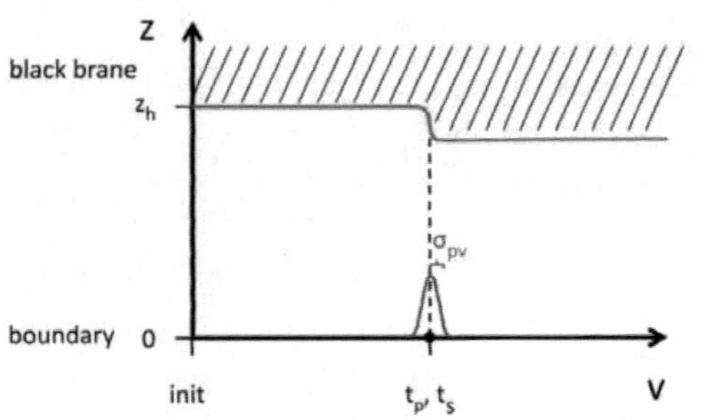


**Figure 2.** Sketch of the spacetime. At boundary time $t_{\mathrm{s}}$ the mass infall reaches its maximum, while $t_{\mathrm{p}}$ defines the time interval of width $\sigma_{\mathrm{pv}}$ at which the metric fluctuation is sourced. The black brane horizon is given by $z_{\mathrm{h}}$ (red line).

## 5. Conclusion

The AdS/CFT correspondence allows the characterization of a plasma far from equilibrium. The position-space Green's function can be a measure of the shear property of the time-dependent plasma. We investigated a 4-dimensional charged asymptotically AdS Vaidya spacetime as a gravitational dual for a homogeneous and isotropic excitation of the plasma. We derived the Green's function in terms of the one-point function in the presence of a Dirac delta-shaped source. Our investigation paves the way for a deeper understanding of the quark-gluon plasma far from equilibrium via the AdS/CFT correspondence.

IOP Conf. Series: Journal of Physics: Conf. Series **942** (2017) 012020 doi:10.1088/1742-6596/942/1/012020

## Acknowledgments

MFW wants to thank the organizers of the *3rd Karl Schwarzschild Meeting on Gravitational Physics and the Gauge/Gravity Correspondence* for an highly interesting and stimulating conference. He is grateful to the Stiftung Polytechnische Gesellschaft Frankfurt am Main. MK was supported in part by the UA "Research in Elementary Particle Physics" DOE grant. The work of PN has been supported by the project "Evaporation of the microscopic black holes" of the German Research Foundation (DFG) under the grant NI 1282/2-2.

## References

[1] Mukhanov V 2012 *Physical Foundations of Cosmology* 3rd ed (New York: Cambridge University Press)

[2] Friman B *et al.* (eds) 2011 *The CBM physics book. Compressed baryonic matter in laboratory experiments* (*Lecture Notes in Physics* vol 814) (Heidelberg: Springer-Verlag)

[3] Ammon M and Erdmenger J 2015 *Gauge/Gravity Duality. Foundations and Applications* (Cambridge: Cambridge University Press)

[4] Gyulassy M and McLerran L 2005 *Nucl. Phys.* A **750** 30–63 (*Preprint* nucl-th/0405013)

[5] Aad G *et al.* (ATLAS) 2010 *Phys. Rev. Lett.* **105** 252303 (*Preprint* 1011.6182)

[6] Kovtun P, Son D T and Starinets A O 2003 *J. High Energy Phys.* JHEP10(2003)064 (*Preprint* hep-th/0309213)

[7] Kovtun P K, Son D T and Starinets A O 2005 *Phys. Rev. Lett.* **94** 111601 (*Preprint* hep-th/0405231)

[8] Buchel A and Liu J T 2004 *Phys. Rev. Lett.* **93** 090602 (*Preprint* hep-th/0311175)

[9] Chesler P M and Yaffe L G 2009 *Phys. Rev. Lett.* **102** 211601 (*Preprint* 0812.2053)

[10] Chesler P M and Yaffe L G 2010 *Phys. Rev.* D **82** 026006 (*Preprint* 0906.4426)

[11] Balasubramanian V *et al.* 2013 *J. High Energy Phys.* JHEP04(2013)069 (*Preprint* 1212.6066)

[12] David J R and Khetrapal S 2015 *J. High Energy Phys.* JHEP07(2015)041 (*Preprint* 1504.04439)

[13] Banerjee S, Ishii T, Joshi L K, Mukhopadhyay A and Ramadevi P 2016 *J. High Energy Phys.* JHEP08(2016)048 (*Preprint* 1603.06935)

[14] Ishii T 2016 Notes on frequencies and timescales in nonequilibrium Green's functions (*Preprint* 1605.08387)

[15] Witten E 1998 *Adv. Theor. Math. Phys.* **2** 505–532 (*Preprint* hep-th/9803131)

[16] Baier R, Romatschke P, Son D T, Starinets O and Stephanov M A 2008 *J. High Energy Phys.* JHEP04(2008)100 (*Preprint* 0712.2451)

[17] Landau L D and Lifshits E M 1966 *Fluid Mechanics* 3rd ed Course of Theoretical Physics (Oxford: Pergamon Press)

[18] Abajo-Arrastia J, Aparício J and López E 2010 *J. High Energy Phys.* JHEP11(2010)149 (*Preprint* 1006.4090)

[19] Caceres E, Kundu A, Pedraza J F and Tangarife W 2014 *J. High Energy Phys.* JHEP01(2014)084 (*Preprint* 1304.3398)

[20] Garfinkle D, Pando Zayas L A and Reichmann D 2012 *J. High Energy Phys.* JHEP02(2012)119 (*Preprint* 1110.5823)

**Institute of Physics**
Dirac House, Temple Back
Bristol BS1 6BE UK

CURRAN ASSOCIATES INC.
proceedings.com

ISSN: 1742-6588
ISBN 978-1-5108-5520-5

9 781510 855205

# 5th International Conference Recent Trends in Structural Materials (COMAT 2018)

IOP Conference Series: Materials Science and Engineering Volume 461

Pilsen, Czech Republic
14 – 16 November 2018

IOP
Institute of Physics